跨境电子商务实训系列

JAVA Chengxu Sheji
Shiyan Jiaocheng

JAVA程序设计
实验教程

许德武 / 主编

浙江大学出版社
ZHEJIANG UNIVERSITY PRESS

图书在版编目(CIP)数据

JAVA 程序设计实验教程/许德武主编.—杭州:浙江大学出版社,2017.7
ISBN 978-7-308-16225-8

Ⅰ.①J… Ⅱ.①许… Ⅲ.①JAVA 语言—程序设计—教材 Ⅳ.①TP312

中国版本图书馆 CIP 数据核字(2016)第 216928 号

JAVA 程序设计实验教程
许德武　主编

丛书策划	朱　玲
丛书主持	曾　熙
责任编辑	曾　熙
责任校对	陈静毅　刘　郡
封面设计	春天书装
出版发行	浙江大学出版社
	(杭州市天目山路 148 号　邮政编码 310007)
	(网址:http://www.zjupress.com)
排　　版	杭州林智广告有限公司
印　　刷	嘉兴华源印刷厂
开　　本	787mm×1092mm　1/16
印　　张	12
字　　数	190 千
版 印 次	2017 年 7 月第 1 版　2017 年 7 月第 1 次印刷
书　　号	ISBN 978-7-308-16225-8
定　　价	28.00 元

版权所有　翻印必究　印装差错　负责调换

浙江大学出版社发行中心邮购电话:(0571)88925591;http://zjdxcbs.tmall.com

跨境电子商务实训系列
编辑委员会

主 任：孙 洁

编 委：潘 蕾　施章清　郑文哲　段文奇

　　　　黄海斌　冯潮前　邹益民　陈素芬

　　　　赫晶晶　曹清玮　徐 燕　许德武

　　　　张俊岭　赵 培　包中文

总　序

跨境电子商务是围绕国家"一带一路""中国制造"等战略的贸易产业新模式，是中国商品实现全球市场"贸易通"的重要路径，是"互联网＋"助力传统贸易转型的具体形式，国务院总理李克强多次强调要大力发展跨境电子商务。当今经济社会，跨境电子商务人才奇缺，优秀的跨境电子商务人才可以说是一将难求。然而，高校在跨境电子商务人才培养方面存在的一个重要问题是缺乏系统性的跨境电子商务系列实训教材，导致高校跨境电子商务实践教学无法满足经济社会的需求。

浙江师范大学文科综合实验教学中心是国家级实验教学示范中心，紧跟国家经济发展战略的重点领域，对接以义乌为中心的浙中区域经济发展特色，在全国领先将跨境电子商务虚拟仿真实验教学作为学校实验教学的重点新兴发展领域，成立了跨境电子商务虚拟仿真实验教学分中心。中心与义乌的中国小商品城集团股份有限公司、阿里巴巴全球速卖通、浙江金义邮政电子商务示范园、金华跨境通等企业开展深度校企合作。中心组织师资团队对跨境电子商务行业领域开展了广泛的调研，明确了跨境电子商务人才所需具备的基本技能与专业技能，并针对这些技能开发跨境电子商务实训系列教材，从而为提高高校跨境电子商务人才培养的教学，尤其是实验教学起到促进作用。

跨境电子商务实训系列教程既可以作为高校电子商务、国际贸易、市场营销等专业的相关实践类课程或理论与实践相结合课程教学的参考教材，也可以作为

跨境电子商务从业人员培训或自学的参考教材。计划出版的跨境电子商务实训系列教程全套共15本，第一期已完成出版的实验教程有7本，分别为：《跨境电子商务平台选择与运营仿真实验教程》（段文奇主编）、《跨境电子商务支付与结算实验教程》（冯潮前主编）、《国际贸易实务仿真模拟实验教程》（徐燕主编）、《物流与供应链虚拟仿真实验》（曹清玮主编）、《电子商务基础实验教程》（黄海滨主编）、《网页设计与制作实验教程》（许德武主编）、《数据库技术与应用实验教程》（张俊岭主编）。第二期将继续推进出版的实验教程有5本，分别为：《电子商务概论》（黄海滨主编）、《电子商务规划与设计实验教程》（吕鑫鑫、李辉、包中文主编）、《跨境电商零创平台实用教程》（邹益民主编）、《JAVA程序设计实验教程》（许德武主编）、《跨境贸易电子商务实操汇编：以金华市为例》（段文奇主编）。

  跨境电子商务实训系列教程的出版是浙江师范大学跨境电子商务虚拟仿真实验教学中心师资团队集体智慧的结晶，本人作为这套系列教程体系的设计者和组织者，对大家的辛勤付出深表敬意。教材出版过程中还得到了浙江师范大学实验室管理处林建军处长、潘蕾副处长，浙江师范大学经济与管理学院郑文哲教授、包中文主任，浙江大学出版社金更达编审、朱玲编辑等出版社工作人员的大力支持，在此一并感谢。

跨境电子商务虚拟仿真实验教学中心主任 孙洁

2015 年 7 月 6 日

前　言

  Java是一门面向对象的编程语言,具有简单性、面向对象、分布式、稳健性、安全性、平台独立与可移植性、多线程、动态性等特点。作为面向对象编程语言的代表,Java极好地实现了面向对象理论,可用于编写桌面应用程序、Web应用程序、分布式系统和嵌入式系统应用程序等。

  本书为Java程序设计上机实验教程。上机实验的目的是提高学生分析问题、解决问题和动手操作的能力。学生通过实践环节理解Java语言的基本结构和程序设计理念,通过亲手编程掌握Java程序设计和编程的方法。

  为了使学生在上机实验时目标明确,本书针对课程内容编写了14个实验。学生可以在课内上机时先完成指导书中给出的程序,理解所学的知识,在此基础上再编写其他应用程序。

  由于作者水平有限,书中难免出现不妥之处,恳请广大读者批评指正。

编　者

2016年10月

目录

实验一　Java 程序运行环境的安装　/ 1

　　一、实验目的　/ 1

　　二、实验要求　/ 1

　　三、实验内容　/ 2

　　四、思考题　/ 14

实验二　Java 语言基础　/ 15

　　一、实验目的　/ 15

　　二、实验要求　/ 15

　　三、实验内容　/ 16

　　四、思考题　/ 25

实验三　类和对象　/ 29

　　一、实验目的　/ 29

　　二、实验要求　/ 29

　　三、实验内容　/ 30

　　四、思考题　/ 49

实验四　数组、向量和字符串　/ 51

　　一、实验目的　/ 51

二、实验要求 / 51

三、实验内容 / 51

四、思考题 / 59

实验五　继承性和多态性　/ 61

一、实验目的 / 61

二、实验要求 / 61

三、实验内容 / 61

四、思考题 / 73

实验六　包、接口与异常处理　/ 75

一、实验目的 / 75

二、实验要求 / 75

三、实验内容 / 75

四、思考题 / 88

实验七　常用系统类的使用　/ 89

一、实验目的 / 89

二、实验要求 / 89

三、实验内容 / 89

四、思考题 / 94

实验八　图形用户界面与多媒体　/ 95

一、实验目的 / 95

二、实验要求 / 95

三、实验内容 / 96

四、思考题 / 122

实验九　流与文件　/ 123

　　一、实验目的　/ 123

　　二、实验要求　/ 123

　　三、实验内容　/ 123

　　四、思考题　/ 136

实验十　线程　/ 137

　　一、实验目的　/ 137

　　二、实验要求　/ 137

　　三、实验内容　/ 137

　　四、思考题　/ 140

实验十一　网络编程　/ 141

　　一、实验目的　/ 141

　　二、实验要求　/ 141

　　三、实验内容　/ 141

　　四、思考题　/ 151

实验十二　数据库的连接：JDBC　/ 153

　　一、实验目的　/ 153

　　二、实验要求　/ 153

　　三、实验内容　/ 153

　　四、思考题　/ 154

实验十三　JSP与Servlet技术　/ 155

　　一、实验目的　/ 155

　　二、实验要求　/ 155

　　三、实验内容　/ 155

　　四、思考题　/ 161

实验十四　综合实验：简单的游戏五子棋　/ 163

　　一、开发环境（实验编译及测试环境）　/ 163

　　二、系统分析　/ 163

　　三、模块功能介绍　/ 174

　　四、功能测试及运行效果　/ 177

　　五、思考题　/ 178

参考文献　/ 179

实验一　Java 程序运行环境的安装

一、实验目的

- 掌握下载 JDK 软件包的方法。
- 掌握设置 Java 程序运行环境的方法。
- 掌握编写与运行 Java 程序的方法。
- 了解 Java 程序设计的基本方法。
- 安装、掌握 Eclipse 软件,为学习 Java 程序设计和进行 Java 程序实验做好准备工作。
- 浏览 Applet 程序。

二、实验要求

- 安装并设置 JDK 软件包。
- 编写简单的 Java 程序。
- 掌握运行 Java 程序的步骤。
- 掌握 Eclipse 软件的安装及配置。

三、实验内容

（一）SDK 的下载与安装

1. 机器要求

Java 对硬件要求不高，下面给出的是基于 Windows 平台的硬件和软件要求。

硬件要求：CPU 奔腾Ⅱ 300M 以上，128M 内存，1.5G 硬盘空间即可。

软件要求：Windows XP/7，IE 8 以上。

2. 下载 SDK

为了建立基于 JDK 的 Java 运行环境，需要先下载 Sun 公司的免费 JDK 软件包。JDK 包含了一整套开发工具，其中包含对编程最有用的 Java 编译器、Applet 查看器和 Java 解释器。

可通过下面的地址下载 JDK 安装程序：http://download.oracle.com/otn-pub/java/jdk/8u60-b27/jdk-8u60-windows-i586.exe? AuthParam＝1442817471_d5d9e5ca2ebbec85403074d988de03c5。

3. 安装 JDK

双击"jdk-8u45-windows-i586.exe"文件开始 JDK 的安装。在安装过程中，注意 JDK 和 JRE 的安装路径(建议采用默认值，如图 1-1、图 1-2 所示)。

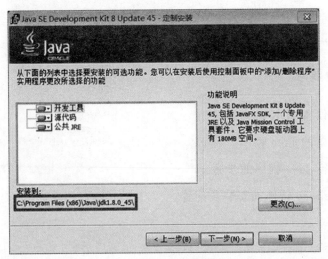

图 1-1 JDK 安装界面

Java程序运行环境的安装 实验一

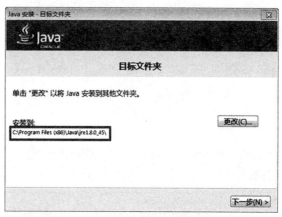

图1-2　JRE安装目标文件夹选择

4. JDK 中的主要文件夹和文件

安装成功后，在安装目录的 bin 文件夹中，包含编译器（javac.exe）、解释器（java.exe）、Applet 查看器（appletviewer.exe）等可执行文件。

（二）设置环境变量

JDK 中的工具都是命令行工具，需要从命令行即 MS-DOS 提示符下运行它们。很多人可能会不习惯，但这是 Sun 特意采取的策略，为的是把精力更多地投入到 Java 程序设计本身而不是工具开发上。

为了能正确方便地使用 JDK，可手工配置 Windows 的环境变量，右击"我的电脑"，选择"属性"，选择"高级"标签，单击"环境变量(N)"按钮，进入环境变量设置界面（如图1-3和图1-4所示）。

图1-3　系统属性界面

图1-4　环境变量设置界面

1. Windows 环境变量手工配置的主要步骤

(1) 创建 JAVA_HOME 变量

单击"新建"按钮,在图 1-5 的红色框中输入 JAVA_HOME 变量值"C:\Program Files (x86)\Java\jdk1.8.0_45"(即 JDK 的安装路径)。效果如图 1-5 所示。

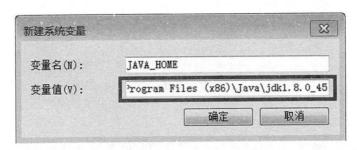

图 1-5 新建 JAVA_HOME 变量界面

(2) 编辑 Path 变量

首先在已有的环境变量中找到并选中 Path 变量(Path 变量一般已经存在),然后单击"编辑"按钮,在原来 Path 变量值的基础上,增加如下内容:;%JAVA_HOME%\bin。效果如图 1-6 所示。

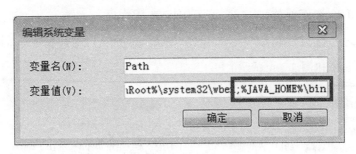

图 1-6 编辑 Path 变量

(3) 创建 CLASSPATH 变量

单击"新建"按钮,在图 1-7 的红色框中输入 CLASSPATH 变量值".;%JAVA_HOME%\lib\dt.jar;%JAVA_HOME%\lib\tools.jar"(注意前面的".;")。效果如图 1-7 所示。

Java程序运行环境的安装 实验一

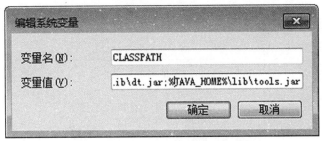

图 1-7　创建 CLASSPATH 变量

(4) 检测 JDK 环境

首先,单击 Windows"开始"菜单,在"搜索程序和文件"输入框中输入"cmd",按 Enter 键进入 MS-DOS 环境。接下来,在 MS-DOS 环境中输入"javac"后按 Enter 键,若能看到如图 1-8 所示界面,说明 JDK 安装配置成功。若显示其他信息,说明 JDK 安装配置不正确,一般是 Path 变量设置不正确,仔细检查并修正第二步的配置错误。

图 1-8　检测 JDK 环境界面

2. Java 语音常用的 MS-DOS 命令

Java 语言中会有一些经常使用的 MS-DOS 命令,如表 1-1 所示。

表 1-1　常用 MS-DOS 命令

序号	命令	实例与含义
1	dir	dir　查看当前目录中的所有子目录和文件 dir *.txt　查看当前目录中的所有文本文件 dir /p　分页查看当前目录中的所有子目录和文件 dir /w　多栏查看当前目录中的所有子目录和文件 dir /?　查看 dir 命令的帮助信息
2	盘符:	d:把当前盘切换到 D 盘 c:把当前盘切换到 C 盘

5

续 表

序号	命令	实例与含义
3	cd	cd c:\windows　把当前目录更改为 c:\windows cd system32　进入当前目录的 system32 子目录 cd boot　进入当前目录的 boot 子目录 cd..　返回父目录 cd\　返回根目录
4	ipconfig	查询本地网络配置
5	ping	ping 192.168.1.1　检测本机是否与 192.168.1.1 相通

(三)编写、编译 Java 源程序

1. 实验内容 1：编译 Helloworld.java 源文件

打开一个纯文本编辑器，键入如下程序：

```
public class Helloworld {
  public static void main(String args[]) {
    System.out.println("Helloworld! ");
  }
}
```

将文件命名为 Helloworld.java，保存为文本格式文件，注意保存文件的路径。根据前面环境变量的设置，Helloworld.java 应该保存在"C:\"路径下。

Java 源程序编写后，要使用 Java 编译器(javac.exe)进行编译，将 Java 源程序编译成可执行的程序代码。Java 源程序都是扩展名为.java 的文本文件。编译时首先读入 Java 源程序，然后进行语法检查，如果出现问题就终止编译。语法检查通过后，生成可执行程序代码即字节码，字节码文件名和源文件名相同，扩展名为.class。编译的参考步骤如图 1-9 所示。

(1) 打开 MS-DOS 环境，并输入"cd \"命令，将当前目录切换到当前盘(C 盘)的根目录。

(2) 输入"javac Helloworld.java"命令，对"Helloworld.java"源文件进行编译，若无错误提示，则表示编译成功。

(3) 输入"dir Helloworld.*"命令，查看文件名为"Helloworld"的所有文件，可以看到编译生成的"Helloworld.class"字节文件。

(4)输入"java Helloworld"命令,运行"Helloworld.class"字节文件。

图 1-9　Helloworld.java 源程序编码界面

2. 实验内容 2:编译 5 层星号金字塔

编写 Java 程序,输出如图 1-10 所示的 5 层星号金字塔。

图 1-10　5 层星号金字塔

(1)首先,使用 JDK 开发的具有图 1-11 所示功能的 Java 源程序,可使用文本编辑器编辑图 1-11 所示的代码,并保存为"c:\PrintDLT.java"。

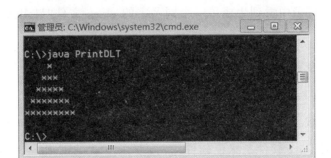

```
public class PrintDLT {
    public static void main(String args[]){
        int i,j;
        for(i=0;i<5;i++){
            for(j=0;j<4-i;j++) System.out.print(" ");
            for(j=0;j<2*i+1;j++) System.out.print("*");
            System.out.println();
        }
    }
}
```

图 1-11　JDK 开发的 Java 源程序

(2) 接下来使用 Java 编译器(javac.exe)对源文件进行编译。参考步骤如下。

①打开 MS-DOS 环境,并输入"cd \"命令,将当前目录切换到当前盘(C 盘)的根目录。

②输入"javac PrintDLT.java"命令,对"PrintDLT.java"源文件进行编译,若无错误提示,则表示编译成功。

③输入"dir PrintDLT.*"命令,查看文件名为"PrintDLT"的所有文件,可以看到编译生成的"PrintDLT.class"字节文件。

(四) 运行 Java 程序

使用 Java 解释器(java.exe)可将编译后的字节码文件 Helloworld.class 解释为本地计算机代码。在命令提示符窗口或 MS-DOS 窗口键入解释器文件名和要解释的字节码文件名 java Helloworld,按 Enter 键即开始解释并可看到运行结果。同样,输入"java PrintDLT"命令,运行"PrintDLT.class"字节文件。

如果源程序没有错误,则屏幕上不显示任何编译错误信息,键入"dir"按 Enter 键后可在目录中看到生成了一个同名字的.class 文件"Helloworld.class"和"PrintDLT.class";否则,将显示出错信息。

(五) 安装、使用 Eclipse 软件

1. 实验内容 1:安装 Eclipse

先上网下载 Eclipse 安装包(下载地址为 http://www.eclipse.org/downloads/),Eclipse 软件是绿色软件,直接解压下载的 Eclipse 安装包即可。解压后的 Eclipse 文件夹内容如图 1-12 所示。

图 1-12 Eclipse 安装包

2. 实验内容 2：使用 Eclipse 开发 Java 程序"Hello world"

（1）启动 Eclipse。双击图 1-12 所示的 eclipse.exe 图标，启动 Eclipse。Eclipse 启动后，会弹出一个如图 1-13 所示的"工作空间启动程序"对话框，工作空间 Workspace 用于保存 Eclipse 所建立的程序项目和相关的设置。在"Workspace"文本框中输入"d：\workspace"目录，单击"OK"按钮。

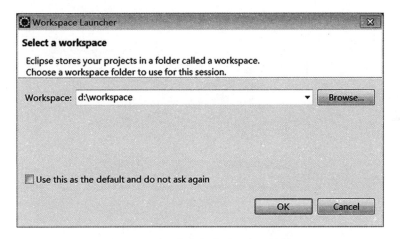

图 1-13 "工作空间启动程序"对话框

单击"OK"按钮后，系统将出现如图 1-14 所示的 Eclipse 欢迎界面。

图 1-14 Eclipse 欢迎界面

(2) 创建 Java 项目。如图 1-15 所示,在 Eclipse 中编写应用程序时,需要先创建一个项目。在 Eclipse 的多种项目中,Java 项目是用于管理和编写 Java 程序的。

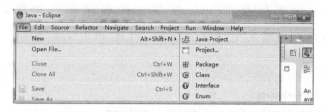

图 1-15 在 Eclipse 中创建 Java 项目文件

执行"File"—"New"—"Java Project"命令,显示如图 1-16 所示的"New Java Project"对话框。

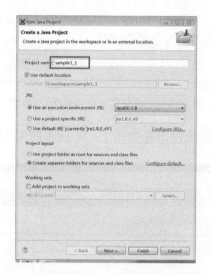

图 1-16 "New Java Project"对话框

在图 1-16 所示的界面中,输入项目名称"sample1_1",然后,单击"Finish"按钮完成项目创建。

(3) 创建 Java 类。在图 1-17 所示的包浏览器(Package Explorer)中选择刚刚创建的项目"sample1_1"并右击,在弹出的快捷菜单中依次选择"New"—"Class"菜单项。系统将显示如图 1-18 所示的"New Java Class"对话框。

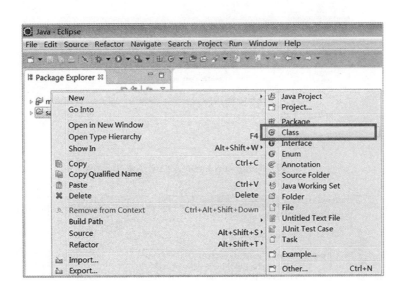

图1-17 包浏览器界面

图1-18 "New Java Class"对话框

在图1-18所示对话框的"Name:"文本框输入Java类名(如Helloworld),并选中下面的"public static void main(String[] args)"复选框后,单击"Finish"按钮。系统将创建Helloworld.java文件,并显示如图1-19所示的界面。

(4) 编写Java类中的main方法。main方法中的代码如图1-20所示。

图1-19 Helloworld.java文件创建成功界面

```
package sample1_1;

public class Helloworld {

    public static void main(String[] args) {
        // TODO Auto-generated method stub
        System.out.println("Hello world!这是我的第一个Java程序");
    }
}
```

图1-20 main方法中的代码

(5)运行Java程序。单击图1-21"Run"—" ▶ "菜单项,或按快捷键Ctrl+F11,即可编译(javac命令)、运行Java程序(java命令)。

图1-21 "Run"菜单项

Eclipse会自动检测代码是否已经保存,如果没有保存,Eclipse会弹出如图1-22所示的对话框,提醒用户对代码进行保存。

单击"OK"按钮,保存代码。Eclipse执行java程序,其运行结果显示在Console(控制台)窗口中。效果如图1-23所示。

图1-22 "提示保存代码"对话框

Java程序运行环境的安装 实验一

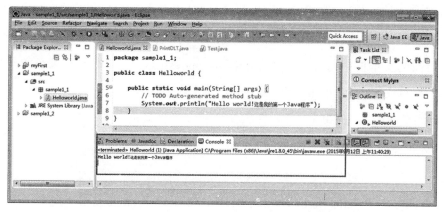

图1-23 Java程序运行结果显示窗口

3. 实验内容3：Eclipse 代码自动补全设置

自动代码补全功能能够帮助我们提高开发效率。例如，当我们忘了某个类的全名，只需要输入开头的几个字母，系统会自动将匹配的类显示出来，供我们选择。要实现这一功能，主要有以下几个步骤。

选择"Windows"—"Preferences"菜单，系统将显示如图1-24所示的参数设置窗口。

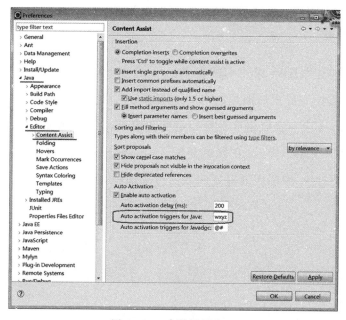

图1-24 参数设置窗口

在图1-24所示窗口左侧的树状列表中依次选择"Java"—"Editor"—"Content Assist",然后在图1-24右面所示的标签"Auto activation triggers for Java"后面的文本框中输入".abcdefgh ijklmnopqrstuvwxyz"(即英文状态的句号+26个小写英文字母)。单击"OK"按钮,关闭当前对话框。

4. 实验内容4:使用Eclipse实现在控制台窗口输出5层星号金字塔

5层星号金字塔具体效果如图1-25所示。

图1-25 5层星号金字塔显示界面

简要的实验步骤如下。

首先,在Eclipse中创建sample1_2项目;接下来,在sample1_2项目中创建一个PrintDLT类;最后,编写该类的main方法,在屏幕上打印5层星号金字塔。

四、思考题

(1) 编写一个"Hello World"的程序。

(2) 设计程序完成矩形面积的求解。要求分两种方式完成:

①在main方法中完成矩形长度和宽度的输入,计算面积后输出。

②先设计一个矩形类,再在测试类的main方法中完成数值(长度和宽度)输入,创建矩形类对象,调用矩形对象的方法计算面积,最后输出结果。

实验二 Java 语言基础

一、实验目的

- 了解 Java 的数据类型。
- 掌握各种变量的声明方式。
- 理解运算符的优先级。
- 掌握 Java 基本数据类型、运算符与表达式。
- 理解 Java 程序语法结构,掌握顺序结构、选择结构和循环结构语法的程序设计方法。
- 通过以上内容,掌握 Java 程序设计的编程规则。

二、实验要求

- 编写一个声明 Java 不同数据类型变量的程序。
- 编写一个使用运算符、表达式、变量的程序。
- 编写一个使用 Java 数组的程序。
- 编写表达式语句、复合语句的程序。
- 编写使用不同选择结构的程序。
- 编写使用不同循环结构的程序。

三、实验内容

（一）声明不同数据类型的变量

1. 编写声明不同数据类型变量的程序文件

编写声明不同数据类型变量的程序文件 KY2_1.java。源代码如下。

```java
public class KY2_1 {
  public static void main(String args[]) {
    byte b=0x55;
    short s=0x55ff;
    int i=1000000;
    long l=0xffffL;
    char c='a';
    float f=0.23F;
    double d=0.7E-3;
    boolean B=true;
    String S="这是字符串类数据类型";
    System.out.println("字节型变量 b = "+b);
    System.out.println("短整型变量 s = "+s);
    System.out.println("整型变量 i = "+i);
    System.out.println("长整型变量 l = "+l);
    System.out.println("字符型变量 c = "+c);
    System.out.println("浮点型变量 f = "+f);
    System.out.println("双精度变量 d = "+d);
    System.out.println("布尔型变量 B = "+B);
    System.out.println("字符串类对象 S = "+S);
  }
}
```

2. 编译并运行程序

编译并运行该程序。

（二）了解变量的使用范围

1. 创建 KY2_2.java 文件

通过本程序了解变量的使用范围,源代码如下。

```
    public class KY2_2 {
      static int i=10;
      public static void main(String args[]) {
        {
          int k=10;
          System.out.println("i="+i);
          System.out.println("k="+k);
        }
        System.out.println("i="+i);
        System.out.println("k="+k);
        //编译时将出错,已超出 k 的使用范围
      }
    }
```

2. 编译 KY2_2.java

此时会出现错误提示。因为变量 k 只在方法块中声明,在方法块之外它是不存在的,所以编译时会出错。

3. 修改上面的程序

根据提示修改程序。

4. 成功运行该程序

修改后运行该程序,直到运行成功。

【请思考】

KY2_2.java 程序说明了什么问题?

(三) 使用关系运算符和逻辑运算符

1. 建立使用关系运算符和逻辑运算符的程序文件

源代码如下。

```
class KY2_3 {
  public static void main(String args[]) {
    int a=25, b=20, e=3, f=0;
    boolean d=a<b;
    System.out.println("a=25,b=20,e=3,f=0");
```

```
            System.out.println("因为关系表达式 a<b 为假,所以其逻辑值为"+d);
        if (e!=0 && a/e>5)
            System.out.println("因为 e 非 0 且 a/e 为 8 大于 5,所以输出 a/e= "+a/e);
        if (f!=0 && a/f>5)
            System.out.println("a/f = "+a/f);
        else
            System.out.println("因为 f 值为 0,所以输出 f = "+f);
    }
}
```

2. 编译并运行该程序

编译并运行该程序直到成功。

(四) 使用表达式语句与复合语句

1. 建立包含表达式语句的程序

源代码如下。

```
class KY2_4{
    public static void main(String[] args) {
        int k, i=3, j=4;
        k=20*8/4+i+j*i;
        System.out.println("表达式(20*8/4+i+j*i)="+k);
    }
}
```

2. 建立包含复合语句的程序

源代码如下。

```
class KY2_5{
    public static void main(String args[]) {
        int k, i=3, j=4;
        k=i+j;
        System.out.println("在复合块外的输出 k="+k);
        {
            float f;
            f=j+4.5F;
            i++;
```

```
            System.out.println("在复合块内的输出 f="+f);
            System.out.println("在复合块内的输出 k="+k);
        }
        System.out.println("在复合块外的输出 i="+i);
    }
}
```

3. 编译并运行上述两个源程序

对源代码进行编译并运行程序。

【请思考】

(1) 将变量 i 定义在块内会怎样？改变其他变量的位置看看会发生什么变化。

(2) 指出程序的复合结构以及变量的使用范围。

(五) 使用选择语句

1. 使用 if...else 语句

(1) 程序功能：使用 if...else 语句构造多分支结构，例如，运用 if...else 语句判断某一年是否为闰年。闰年的条件是符合下面两个条件之一：能被 4 整除，但不能被 100 整除；能被 400 整除。

(2) 编写源程序文件，代码如下。

```
public class KY2_6 {
    public static void main(String args[]) {
        boolean leap;
        int year=2005;
        if ((year%4==0 && year%100!=0) || (year%400==0))  //方法 1
            System.out.println(year+"年是闰年");
        else
            System.out.println(year+"年不是闰年");
        year=2008;  //方法 2
        if (year%4!=0)
            leap=false;
        else if (year%100!=0)
            leap=true;
        else if (year%400!=0)
```

            leap=false;
          else
            leap=true;
          if(leap==true)
            System.out.println(year+"年是闰年");
          else
            System.out.println(year+"年不是闰年");
          year=2050;//方法 3
          if(year%4==0){
            if(year%100==0){
              if(year%400==0)
                leap=true;
              else
                leap=false;
            }
            else
              leap=false;
          }
          else
            leap=false;
          if(leap==true)
            System.out.println(year+"年是闰年");
          else
            System.out.println(year+"年不是闰年");
        }
      }

（3）编译运行程序。

【请思考】

本程序中有几个选择语句,哪些具有嵌套关系?

2.使用 switch 语句

（1）程序功能:在不同温度时显示不同的解释说明。

(2) 程序源代码如下。

```java
class KY2_7{
    public static void main(String args[]) {
        int c=38;
        switch (c<10? 1: c<25? 2: c<35? 3: 4) {
            case 1:
                System.out.println(""+c+"℃ 有点冷。要多穿衣服。");
            case 2:
                System.out.println(""+c+"℃ 正合适。出去玩吧。");
            case 3:
                System.out.println(""+c+"℃ 有点热。");
            default:
                System.out.println(""+c+"℃ 太热了！开空调。");
        }
    }
}
```

(3) 编译并运行程序。

(六) 使用循环语句

1. for 循环语句练习

(1) 程序功能：按 5 摄氏度的增量打印出一个从摄氏温度到华氏温度的转换表。

(2) 程序源代码如下。

```java
class KY2_8{
    public static void main (String args[]) {
        int h,c;
        System.out.println("摄氏温度 华氏温度");
        for (c=0; c<=40; c+=5) {
            h=c*9/5+32;
            System.out.println(" "+c+"          "+h);
        }
    }
}
```

(3) 编译并运行程序。

2. while 循环语句练习

(1) 程序功能：运行程序后从键盘输入数字 1、2、3 后，可显示抽奖得到的奖品；如果输入其他数字或字符则显示"真不幸，你没有奖品！下次再来吧。"。

(2) 程序源代码如下。

```java
import java.io.*;
class KY2_9 {
  public static void main(String args[]) throws IOException {
    char ch;
    System.out.println("按 1、2、3 数字键可得大奖！");
    System.out.println("按空格键后回车可退出循环操作。");
    while ((ch=(char)System.in.read())!=' ')
    {
      System.in.skip(2);      //跳过回车键
      switch(ch) {
        case '1':
          System.out.println("恭喜你获得大奖，一辆汽车！");
          break;
        case '2':
          System.out.println("不错呀，你得到一台笔记本电脑！");
          break;
        case '3':
          System.out.println("没有白来，你得到一台冰箱！");
          break;
        default:
          System.out.println("真不幸，你没有奖品！下次再来吧。");
      }
    }
  }
}
```

(3) 编译源程序。

(4) 在命令提示符窗口运行程序,然后分别输入 1、2、3、r ,记录结果。

3. do...while 循环语句练习

(1) 程序功能：求 1+2+…+100 之和,并将求和表达式与所求的和显示出来。

(2) 程序源代码如下。

```
class KY2_10 {
  public static void main(String args[]) {
    int n=1, sum=0;
    do {
      sum+=n++;
    }
    while (n<=100);
    System.out.println("1+2+...+100 ="+sum);
  }
}
```

(3) 编译并运行程序。

4. 多重循环练习

(1) 输出九九乘法表的程序,源代码如下。

```
public class KY2_11
{
  public static void main(String args[])
  {
    int i,j,n=9;
    System.out.print("    *    |");
    for (i=1;i<=n;i++)
      System.out.print("  "+i);
    System.out.print("\n———————|");
    for (i=1;i<=n;i++)
      System.out.print("————");
    System.out.println();
    for (i=1;i<=n;i++)
    {
```

```
            System.out.print("    "+i+"    |");
            for(j=1;j<=i;j++)
              System.out.print(" "+i*j);
            System.out.println();
          }
        }
      }
```

(2) 编译并运行程序。

(七) 综合实例

1. 计算学分绩点

使用 Eclipse 创建 Java 项目"sample2_1",然后为该项目创建 Java 类 "CalcXFJD",并在该类的 main 方法中实现"计算学分绩点"功能。具体步骤如下：输入一个百分制成绩(若输入的成绩大于100,则要求用户重新输入),根据输入的成绩计算其学分绩点(计算公式参见表2-1)。程序可多次计算学分绩点,直到输入的成绩为负数。具体效果如图2-1所示。具体运行功能请参考样例程序 CalcXFJD.class(先下载样例程序到 C:\,然后进入 MS-DOS 环境,把当前目录切换到 C:\,再通过命令"C:\>java CalcXFJD"运行该程序,并体会其功能)。

表 2-1 学分绩点计算公式

成绩(score)	绩点
score=100	5.0
95≤score<100	4.5
90≤score<95	4.0
85≤score<90	3.5
80≤score<85	3.0
75≤score<80	2.5
70≤score<75	2.0
65≤score<70	1.5
60≤score<65	1.0
score<60	0.0

图 2-1 输入成绩计算绩点界面

2. 查找质数并显示

使用 Eclipse 创建 Java 项目"sample2_2",然后为该项目创建 Java 类"FindPrimes",并在该类的 main 方法中实现"查找质数并输出"功能。具体功能如下:找出 100~500 所有的质数,并以 5 个一行显示,具体效果如图 2-2 所示。具体运行功能请参考样例程序 FindPrimes.class(先下载样例程序到 C:\,然后进入 MS-DOS 环境,把当前目录切换到 C:\,再通过命令"C:\>java FindPrimes"运行该程序,并体会其功能)。

图 2-2 查找并显示 100~500 的质数界面

3. 显示某月日历

使用 Eclipse 创建 Java 项目"sample2_3",然后为该项目创建 Java 类"MyCalendar",并在该类的 main 方法中实现"显示某月日历"功能。具体功能如下:输入不小于 1900 的年份(若输入的年份不符合要求,则要求重新输入年份),再输入月份(若月份不符合要求,则要求重新输入月份),程序显示指定年份指定月份的日历。具体效果如图 2-3 所示。具体运行功能也可参考样例程序 MyCalendar.class(先下载样例程序到 C:\,然后进入 MS-DOS 环境,把当前目录切换到 C:\,再通过命令"C:\>java MyCalendar"运行该程序,并体会其功能)。

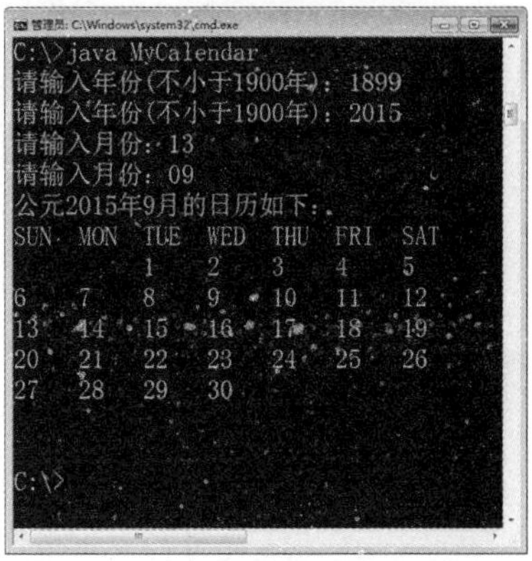

图 2-3 创建并显示日历界面

四、思考题

(1) 求任意一个给定整数的反序数。反序数就是将整数各个位置上的数字顺序倒过来形成的整数。如 12345 的反序数是 54321。

(2) 求解百鸡问题(出自《张丘建算经》):鸡翁一,值钱五;鸡母一,值钱三;鸡雏三,值钱一。百钱买百鸡,问鸡翁、鸡母、鸡雏各几何?

(3) 打印图形(如打印正立/倒立等腰三角形、菱形、空心菱形等)。

(4) (选做)判断用户输入的数是否为水仙花数。水仙花数是指一个 n 位(n≥

3)的数字,它的每个位上的数字的 n 次幂之和等于它本身。例如,1×1×1 + 5×5×5 + 3×3×3 = 153。

(5) 设计一个程序完成对彩票中奖情况的模拟。根据彩票的末尾号码来判断是否中奖,末尾号码为 1、3、9 是三等奖;末尾后 2 位为 29、49、99 是二等奖;末三位为 123、128、789 是一等奖。实现时要求设计两个类 Guess 和 GuessTest。Guess 完成是否中奖以及中几等奖的判断,GuessTest 完成彩票号码的产生及结果输出。用 if 语句或 switch 语句实现。

(6) 编写程序求解 1 到 100 的累加和,并从控制台输出。

(7) 分析下面的程序,说出下面的程序为什么是死循环?

```java
class Sum {
    public static void main(String args[]) {
        int i=1,n=10,s=0;
        while (i<=n)
        s = s + i;
        System.out.println("s="+s);
    }
}
```

(8) 分析下面源程序的结构,写出运行结果。

```java
class CircleArea {
    final static double PI=3.14159;
    public static void main(String args[]) {
        double r1=8.0, r2=5.0;
        System.out.println("半径为"+r1+"的圆面积="+area(r1));
        System.out.println("半径为"+r2+"的圆面积="+area(r2));
    }
    static double area(double r) {
        return (PI * r * r);
    }
}
```

(9) 编写程序,根据考试成绩的等级(A、B、C、D、E)打印出百分制分数段。将 A 设为 90 分及以上,B 设为 90 分以下 80 分及以上,C 设为 80 分以下 70 分及以上,D 设为 70 分以下 60 分及以上,E 设为 60 分以下。要求在程序中使用 switch 语句。

（10）将下面的程序补充完整,利用 break 语句和带标号的 break 语句分别退出一重循环和二重循环。

```
for (i=0; i<10; i++){
  int j=i*10
  while(j<100){
    if (j==10) break;
    j=j+5;
  }
}
```

实验三 类和对象

一、实验目的

- 理解 Java 程序设计是如何体现面向对象编程的基本思想的。
- 了解类的封装方法,以及如何创建类和对象。
- 了解成员变量和成员方法的特性,掌握 OOP 方式进行程序设计的方法。
- 了解类的继承性和多态性的作用。

二、实验要求

- 编写一个体现面向对象思想的程序。
- 编写一个创建对象和使用对象的方法的程序。
- 编写一个显示当前日期和时间的程序。
- 编写不同成员变量修饰方法的程序。
- 编写不同成员方法修饰方法的程序。
- 编写体现类的继承性(成员变量、成员方法、成员变量隐藏)的程序。
- 编写体现类的多态性(成员方法重载、构造方法重载)的程序。
- 理解和掌握成员方法中的参数传递规则。

三、实验内容

（一）类的声明与使用

1. 使用 Eclipse 创建名为"task3_1"的 Java 项目

在该项目中创建一个名为"Rectangle"的 Java 类，该类实现图 3-1 所示 UML 类图的功能（Rectangle 类代码自行实现）。在"task3_1"项目中再创建一个名为"Task3_1"的 Java 主类，在该类的 main 方法中使用 Rectangle 类，参考代码如图 3-2 所示。

```
Rectangle
width : double
height : double
getArea() : double
```

图 3-1　Rectangle UML 类图功能界面

```java
public static void main(String[] args) {
    Rectangle myRect1;                          //声明Rectangle类变量myRect1
    myRect1 = new Rectangle();                  //创建Rectangle对象并赋值给变量myRect1
    myRect1.width = 10;                         //访问对象myRect1的成员变量
    myRect1.height = 20;
    double area = myRect1.getArea();            //访问对象myRect1的成员方法
    System.out.println(area);
}
```

图 3-2　main 方法中使用 Rectangle 类参考代码

2. 使用 Eclipse 创建名为"task3_2"的 Java 项目

在该项目中创建一个名为"Sector"的 Java 类，该类实现图 3-3 所示 UML 类图的功能。在"task3_2"项目中再创建一个名为"Task3_2"的 Java 主类，在该类的 main 方法中使用 Sector 类，参考代码如图 3-4 所示。

```
Sector
radius : double
startAngle : double
endAngle : double
getArea() : double
getPerimeter() : double
```

图 3-3　Sector UML 类图功能界面

```
public static void main(String[] args) {
    Sector mySector;                          //声明Sector类型的变量mySector
    mySector = new Sector();                  //创建Sector类型的对象并赋值给变量mySector
    mySector.radius = 10;                     //访问对象mySector的成员对象
    mySector.startAngle = 0;
    mySector.endAngle = 1.57;
    double s = mySector.getArea();            //访问对象mySector的成员方法
    System.out.println(s);
}
```

图 3-4　main 方法中使用 Sector 类参考代码

(二) 方法的重载

使用 Eclipse 创建名为"task3_3"的 Java 项目,在该项目中创建一个名为"Maths"的 Java 类,先定义一个 max 方法,用于实现求两个整数中的最大值。再定义一个 max 重载方法,它实现求三个整数中的最大值。在"task3_3"项目中再创建一个名为"Task3_3"的 Java 主类,在该类的 main 方法中使用 Maths 类,参考代码如图 3-5 所示。

```
public static void main(String args[]){
    int x,y,z;
    Scanner reader = new Scanner(System.in);
    System.out.println("请输入变量x的值: ");
    x = reader.nextInt();
    System.out.println("请输入变量y的值: ");
    y = reader.nextInt();
    System.out.println("请输入变量z的值: ");
    z = reader.nextInt();
    reader.close();
    Maths myMaths = new Maths();
    System.out.println("x和y中的最大值是"+myMaths.max(x, y));
    System.out.println("x,y和z中的最大值是"+myMaths.max(x, y,z));
}
```

图 3-5　main 方法中使用 Maths 类参考代码

(三) 类的构造方法

使用 Eclipse 创建名为"task3_4"的 Java 项目,在该项目中创建一个名为"Rectangle"的 Java 类,再实现图 3-6 所示 UML 类图。其中包含两个构造方法:一个是无参数的构造方法,将成员变量 width 和 height 都设置为 1;另一个是带参数 w 和 h 的构造方法,将成员变量 width 和 height 分别设置为参数 w 和 h。在"task3_4"项目中再创建一个名为"Task3_4"的 Java 主类,在该类的 main 方法中使用 Rectangle 类的两个构造方法,参考代码如图 3-7 所示。

图 3-6 Rectangle UML 类图(含两个构造方法)

```java
public static void main(String[] args) {
    Rectangle myRect1, myRect2;              //声明Rectangle类变量myRect1
    myRect1 = new Rectangle();               //调用第1种构造方法（即无参数的构造方法）
    myRect1.width = 10;                      //设置width成员变量
    myRect2 = new Rectangle(4, 5);           //调用第2种构造方法（即有参数的构造方法）
    System.out.println(myRect1.getArea());
    System.out.println(myRect2.getArea());
}
```

图 3-7 main 方法中使用 Rectangle 类的两个构造方法的参考代码

（四）对象的引用和实体

1. 实验内容 1：阅读程序，理解对象的引用和实体，写出程序运行输出结果

已知 Rectangle 类(定义在文件 Rectangle.java 中)的代码如图 3-8 所示，主类 Task3_5 的 main 方法(定义在文件 Task3_5.java 中)的代码如图 3-9 所示。

```java
public class Rectangle {                     //定义一个Rectangle类
    double width;                            //成员变量width表示矩形的宽度
    double height;                           //成员变量height表示矩形的高度
    double getArea(){                        //成员方法getArea()用来计算矩形的面积
        return width * height;
    }
    Rectangle(double w, double h) {
        width = w;
        height = h;
    }
    Rectangle() {
        width=1; height=1;
    }
}
```

图 3-8 Rectangle.java 文件代码

```java
public static void main(String[] args) {
    Rectangle myRect1, myRect2;              //声明Rectangle类变量myRect1
    myRect1 = new Rectangle();               //调用第1种构造方法（即无参数的构造方法）
    myRect1.width = 10;                      //设置width成员变量
    myRect2 = new Rectangle(4, 5);           //调用第2种构造方法（即有参数的构造方法）
    System.out.println(myRect1.getArea());
    System.out.println(myRect2.getArea());
    myRect1 = myRect2;
    System.out.println(myRect1.getArea());
}
```

图 3-9 Task3_5.java 文件代码

2. 实验内容2：在Eclipse中编写上述代码，验证阅读程序所得到结果

使用Eclipse创建名为"task3_5"的Java项目，在该项目中创建一个名为"Rectangle"的Java类，代码如图3-8所示。在"task3_5"项目中再创建一个名为"Task3_5"的Java主类，该类的main方法代码如图3-9所示。

（五）参数传递

1. 实验内容1：阅读程序，理解参数的传递规则，写出其运行输出结果

已知Battery类、Radio类和MyMath类（这三个类都定义在文件Battery.java中）的代码如图3-10所示，主类Task3_6的main方法（定义在文件Task3_6.java中）的代码如图3-11所示。

```java
public class Battery {                  //Battery类表示电池
    double electricityAmount;
    public Battery(double eAmount) {
        electricityAmount = eAmount;
    }
    public void comsumePower(double c){
        electricityAmount -= c;
    }
}
class Radio{                            //Radio类表示收音
    public void openRadio(Battery battery){  //形参battery属于引用类型
        battery.comsumePower(10);
    }
}
class MyMath{       //MyMath类定义了一些常用的数学计算
    public int calc(int n){             //形参n属于基本类型
        int s = 0;
        while(n>0){
            int c = n % 10; s += c; n = n / 10;
        }
        System.out.println("形参n的值是"+n);
        return s;
    }
}
```

图3-10 Battery.java文件代码

```java
public static void main(String[] args) {
    int x = 13579;
    MyMath mathTool = new MyMath();
    int y = mathTool.calc(x);                       //实参x属于基本类型
    System.out.println(x+"的各数位之和为"+y);
    Battery nanfu = new Battery(100);
    System.out.println("nanfu电池的当前电量为"+nanfu.electricityAmount);
    Radio myRadio = new Radio();
    System.out.println("使用nanfu电池来打开收音机");
    myRadio.openRadio(nanfu);                       //实参nanfu属于引用类型
    System.out.println("nanfu电池的当前电量为"+nanfu.electricityAmount);
}
```

图3-11 Task3_6.java文件代码

2. 实验内容 2：在 Eclipse 中编写上述代码，验证阅读程序所得到结果

使用 Eclipse 创建名为"task3_6"的 Java 项目，在该项目中创建三个名称分别为"Battery""Radio"和"MyMath"的 Java 类，代码如图 3-10 所示（这三个类都定义在文件 Battery.java 中）。在"task3_6"项目中再创建一个名为"Task3_6"的 Java 主类，该类的 main 方法代码如图 3-11 所示。

3. 实验内容 3：编写一个传值调用的程序文件 task3_7.java

(1) 程序功能：首先给整型变量 x 和 y 赋一个初值 10，然后使用传值调用方式调用方法 ff1 对 x 和 y 做乘方及输出 x 和 y 的乘方值，最后再输出 x 和 y 的乘方值。

(2) 程序源代码如下。

```
class task3_7 {
    public static void main(String[] args) {
        int x=10, y=10;
        ff1(x, y);
        System.out.println("x="+x+", y="+y);
    }
    static void ff1(int passX, int passY) {
        passX=passX * passX;
        passY=passY * passY;
        System.out.println("passX="+passX+", passY="+passY);
    }
}
```

(3) 编译 task3_7.java。

(4) 分析其运行结果。

4. 实验内容 4：编写一个调用对象方法的程序文件 task3_8.java

(1) 程序功能：通过调用对象的方法在方法调用后修改成员变量的值。

(2) task3_8.java 程序源代码如下。

```
class task3_8 {
    public static void main(String[] args) {
        Power p=new Power();
```

```
      p.ff2(10,10);
      System.out.println("方法调用后 x="+p.x+", y="+p.y);
    }
  }
  class Power{
    int x=10, y=10;
    void ff2(int passX, int passY) {
      System.out.println("初始时 x="+x+", y="+y);
      x=passX * passX;
      y=passY * passY;
      System.out.println("方法调用中 x="+x+", y="+y);
    }
  }
```

（3）编译 task3_8.java。

以上实验例子仅仅是为了说明 Java 编程中参数传递时要注意的问题，在实际编程中是不可取的，因为完全可以采用其他更好的方法来实现参数的传递。例如，前面还使用过传递对象的方式。

【请思考】

方法的参数传递有哪些方式？区别是什么？

（六）对象的组合

1. 实验内容1：掌握在 Eclipse 中实验对象的组合方法，理解和掌握对象的关联关系和依赖关系

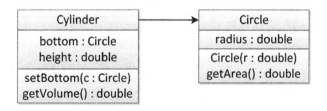

图 3-12　task3_9 UML 类图

使用 Eclipse 创建名为"task3_9"的 Java 项目，在该项目中创建一个名称为"Circle"的 Java 类（表示圆），其成员变量和方法如图 3-12 右侧所示。在

"task3_9"项目中再创建一个名称为"Cylinder"的Java类(表示圆柱体),其成员变量和方法如图3-12左侧所示,其中,成员变量bottom为Circle类的对象。在"task3_9"项目中再创建一个名为"Task3_9"的Java主类,该类的main方法代码如图3-13所示。

```
public static void main(String[] args) {
    Circle c1 = new Circle();              //创建一个圆对象c1
    c1.radius = 10;                        //设置圆的半径
    Cylinder myCylinder = new Cylinder();  //创建一个圆柱体对象
    myCylinder.setBottom(c1);              //设置圆柱体的底圆为c1
    myCylinder.height = 50;                //设置圆柱体的高
    System.out.println(myCylinder.getVolume()); //输出圆柱体的体积
}
```

图3-13　Task3_9中的Java代码

2. 实验内容2:判断下面两个UML类图中类与类之间的关系

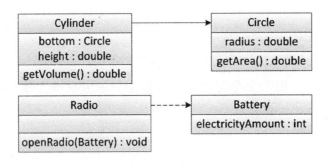

图3-14　判断类与类的关系

【问题1】在图3-14的UML类图中,Cylinder类与Circle类是什么关系?

回答:(　　)类(关联或依赖)(　　)类。

【问题2】在图3-14的UML类图中,Radio类与Battery类是什么关系?

回答:(　　)类(关联或依赖)(　　)类。

(七) 实例成员与类成员

1. 实验内容1:阅读程序,理解实例成员和类成员之间的区别,写出程序运行输出结果

已知Account类(定义在文件Account.java中)的代码如图3-15所示,主类Task3_10的main方法(定义在文件Task3_10.java中)的代码如图3-16所示。

```
//Account类表示银行账户
public class Account {
    static double totalMoney;        //类成员变量,表示银行总资金量
    String accountID;                //实例成员变量,表示账户ID
    double money;                    //实例成员变量,表示账户金额
    Account(String id) {             //构造方法,参数id用于设置账户ID
        accountID = id;
    }
}
```

图 3-15　Account.java 文件代码

```
public static void main(String[] args) {
    Account.totalMoney = 1000000;          //通过类名访问类成员变量
    Account a1 = new Account("001");
    a1.money += 1000;                      //必须通过对象访问实例成员变量
    a1.totalMoney += 1000;                 //也可以通过对象访问类成员变量
    Account a2 = new Account("002");
    a2.money += 2000;
    a2.totalMoney += 2000;                 //也可以通过对象访问类成员变量
    System.out.println("账户"+a1.accountID+"有"+a1.money+"元");
    System.out.println("账户"+a2.accountID+"有"+a2.money+"元");
    //下面的输出说明Account类的所有对象共享类成员变量totalMoney
    System.out.println("目前,银行共有资金"+Account.totalMoney+"元");
}
```

图 3-16　Task3_10.java 文件代码

2. 实验内容 2：在 Eclipse 中编写上述代码，验证阅读程序所得到结果

使用 Eclipse 创建名为"task3_10"的 Java 项目，在该项目中创建"Account"的 Java 类，代码如图 3-15 所示（定义在文件 Account.java 中）。在"task3_10"项目中再创建一个名为"Task3_10"的 Java 主类，该类的 main 方法代码如图 3-16 所示。

（八）实例方法与类方法

1. 实验内容 1：阅读程序，理解实例方法和类方法之间的区别，写出程序运行输出结果

已知 Account 类（定义在文件 Account.java 中）的代码如图 3-17 所示，主类 Task3_11 的 main 方法（定义在文件 Task3_11.java 中）的代码如图 3-18 所示。

```
public class Account {                              //银行账户类
    static double totalMoney;                       //类(或静态)成员变量
    String accountID;                               //实例成员变量
    double money;                                   //实例成员变量
    Account(String id) {
        accountID = id;
    }
    static void addMoney(double am){                //类方法,向银行注入资金
        totalMoney += am;                           //在类方法中访问类成员变量
        System.out.println("该银行注入资金"+am+"元");
    }
    static void showBankMoney(){                    //类方法,显示银行资金总量
        System.out.println("该银行资金量为"+totalMoney+"元");
    }
    void showAccountMoney(){
        System.out.println("账户"+accountID+"有资金"+money+"元");
        showBankMoney();                            //实例方法中调用类方法
    }
    double deposit(double m){                       //实例方法,存钱
        money += m;                                 //在实例方法中访问实例成员变量
        totalMoney += m;                            //在实例方法中访问类成员变量
        System.out.println("账户"+accountID+"存入了"+m+"元");
        return money;
    }
    double withdraw(double m){                      //实例方法,取钱
        money -= m;                                 //在实例方法中访问实例成员变量
        totalMoney -= m;                            //在实例方法中访问类成员变量
        System.out.println("账户"+accountID+"取出了"+m+"元");
        return m;
    }
}
```

图 3-17 Account.java 文件代码

```
public static void main(String[] args) {
    Account.addMoney(1000000);                      //通过类名访问类方法
    Account account1 = new Account("001");
    account1.deposit(2000);                         //通过对象访问实例方法
    account1.addMoney(2000000);                     //通过对象访问类方法
    Account.showBankMoney();                        //通过类名访问类方法
    account1.withdraw(500);
    account1.showAccountMoney();                    //通过对象访问实例方法
}
```

图 3-18 Task3_11.java 文件代码

2. 实验内容 2：回答下面 9 个判断题

(1) 用关键字 static 只能修饰成员变量,不能修饰成员方法。（   ）

(2) 静态(类)方法和静态(类)成员只能通过类名进行访问。（   ）

(3) 实例方法和实例成员变量可以通过类名进行访问。（   ）

(4) 静态(类)方法只能访问静态(类)成员变量和其他静态(类)方法。（   ）

(5) 实例方法可以访问静态(类)方法和静态(类)成员。（   ）

(6) 同一类的不同对象的实例成员变量相互独立。（   ）

(7) 同一类的不同对象共享其类成员变量。（   ）

(8) 必须创建该类的一个对象后,才可以访问该类的静态成员变量。(　　)

(9) 在静态(类)方法中可以使用关键字 this。(　　)

3. 实验内容3:在 Eclipse 中编写上述代码,验证阅读程序所得到结果

使用 Eclipse 创建名为"task3_11"的 Java 项目,在该项目中创建"Account"的 Java 类,代码如图 3-17 所示(定义在文件 Account.java 中)。在"task3_11"项目中再创建一个名为"Task3_11"的 Java 主类,该类的 main 方法代码如图 3-18 所示。

(九) 在 JDK 中使用包

1. 实验内容1:理解包的含义,掌握在 JDK 中使用包的方法

(1) 步骤1:采用记事本编辑 Circle 类,代码如图 3-19 所示。

```
package aaa.bbb;                    //指定包名
public class Circle{
    public double radius;
    public double getArea(){
        System.out.println("aaa.bbb.Circle");
        return radius * radius * 3.14;
    }
}
```

图 3-19　Circle 类代码

(2) 步骤2:保存指定包名的类。首先在 D 盘的根目录中新建文件夹"D:\task3_12",并在 D:\task3_12 文件夹下创建"aaa"文件夹,再在"aaa"文件夹下创建"bbb"文件夹,再将上面编辑的 Circle 类保存到文件"D:\task4_10\aaa\bbb\Circle.java"。

注意:指定包名的类的存放路径必须与包名相一致,即如果某个类指定的包名为 xxx.yyy.zzz,则该类必须保存在"<应用程序根目录>\xxx\yyy\zzz"文件夹中。

(3) 步骤3:编译指定了包名的 Circle 类。打开 MS-DOS 环境,通过 cd 命令把当前目录切换到"D:\task3_12",再通过如图 3-20 所示命令对文件 D:\task3_12\aaa\bbb\Circle.java 进行编译。

```
D:\task4_10>javac aaa\bbb\circle.java
```

图 3-20　编译指定了包名的 Circle 类命令

(4) 步骤4:采用记事本编辑 Task3_12 类,采用第1种方法使用 aaa.bbb 包中的 Circle 类,代码如图3-21所示。将 Task3_12 类保存到文件"D:\task3_12\Task3_12.java"。

```
public class Task3_12{
    public static void main(String args[]){
        //The first method to use class Circle defined in package aaa.bbb
        aaa.bbb.Circle c1 = new aaa.bbb.Circle();
        c1.radius = 10;
        System.out.println(c1.getArea());
    }
}
```

图3-21 记事本编辑 Task3_12 类

(5) 步骤5:编译 Task3_12 类。打开 MS-DOS 环境,通过 cd 命令把当前目录切换到"D:\task3_12",再通过如图3-22所示命令对文件 D:\task3_12\Task3_12.java 进行编译。

```
D:\task3_12>javac task3_12.java
```

图3-22 编译 Task3_12 类命令

(6) 步骤6:采用记事本编辑 Task4_10A 类,采用第2种方法使用 aaa.bbb 包中的 Circle 类,代码如图3-23所示。将 Task3_12A 类保存到文件"D:\task3_12\Task3_12A.java"。

```
//The second method to use class Circle defined in package aaa.bbb
import aaa.bbb.Circle; //import the class Circle defined in package aaa.bbb
public class Task4_10A{
    public static void main(String args[]){
        Circle c1 = new Circle();   // use the class Circle directly
        c1.radius = 10;
        System.out.println(c1.getArea());
    }
}
```

图3-23 记事本编辑 Task3_12A 类

(7) 步骤7:编译 Task3_12A 类。打开 MS-DOS 环境,通过 cd 命令把当前目录切换到"D:\task3_12",再通过如图3-24所示命令对文件 D:\task3_12\Task3_12A.java 进行编译。

```
D:\task3_12>javac task3_12A.java
```

图3-24 编译 Task3_12A 类命令

(8) 步骤 8：执行 Task3_12.class 和 Task3_12A.class。打开 MS-DOS 环境，通过 cd 命令把当前目录切换到"D:\task3_12"，再通过如图 3-25 和图 3-26 所示命令分别执行 Task3_12.class 和 Task3_12A.class。

图 3-25　Task3_12.class 命令

图 3-26　Task3_12A.class 命令

（十）在 Eclipse 中使用包

1. 实验内容 1：在 Eclipse 中使用包

使用 Eclipse 创建 Java 项目"task3_13"，在该项目中创建一个名为"MyCircle"的 Java 类，在"New Java Class"对话框中指定 MyCircle 类的包名，效果如图 3-27 所示。MyCircle 类的代码如图 3-28 所示。

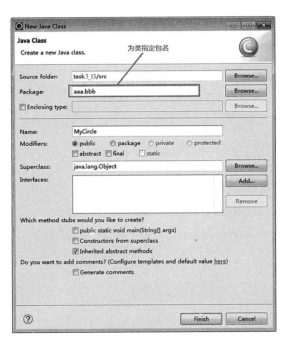

图 3-27　在"New Java Class"对话框中指定 MyCircle 类包名的界面

```
package aaa.bbb;
public class MyCircle {
    public double r;
    public double getArea(){
        System.out.println("aaa.bbb.MyCircle");
        return r * r * 3.14;
    }
}
```

图 3-28　MyCircle 类代码

在项目 task3_13 中创建一个名为"Task3_13"的 Java 类,在"New Java Class"对话框中不指定 Task3_13 类的包名,效果如图 3-29 所示。Task3_13.java 文件中的代码如图 3-30 所示(包含了 MyCircle 类和 Task3_13 类的代码)。

图 3-29　"New Java Class"对话框中不指定 Task3_13 类包名的界面

```
class MyCircle{
    double radius;
    public MyCircle(double r) {
        radius = r;
    }
    public MyCircle(){
    }
    double getArea(){
        System.out.println("同一包中的MyCircle");
        return radius * radius * 3.14;
    }
}
public class Task3_13 {
    public static void main(String[] args) {
        //使用同一包中的MyCircle类
        MyCircle c1 = new MyCircle(10);
        System.out.println(c1.getArea());
        //使用包aaa.bbb中的MyCircle类
        aaa.bbb.MyCircle c2 = new aaa.bbb.MyCircle();
        c2.r = 10;
        System.out.println(c2.getArea());
    }
}
```

图 3-30　Task3_13.java 文件代码

(十一) 类成员的访问权限

1. 实验内容1：通过实验理解类成员的各种访问权限

使用 Eclipse 创建 Java 项目"task3_14"，在该项目中创建一个名为"AccessControl"的 Java 类，在"New Java Class"对话框中指定 AccessControl 类的包名为"task3_14.package1"，效果如图 3-31 所示。AccessControl 类的代码如图 3-32 所示。

图 3-31　指定包名为 task3_14.package1 对话框

```
//AccessControl类存放在包task3_14.package1中
package task3_14.package1;

public class AccessControl {
    public int variable1;      //公共变量
    int variable2;             //友好变量
    public int method1(){      //公共方法
        return 1;
    }
    int method2(){             //友好方法
        return 2;
    }
}
```

图 3-32 AccessControl 类代码

在项目 task3_14 中创建一个名为"Task3_14"的 Java 类,在"New Java Class"对话框中指定 Task3_14 类的包名为"task3_14",效果如图 3-33 所示。Task3_14.java 文件中的代码如图 3-34 所示(包含了 Task3_14 类和另一个 AccessControl 类的代码)。

图 3-33 在"New Java Class"对话框中指定 Task3_14 类的包名界面

```
package task3_14;
class AccessControl {
    public int variable1;      //公共变量
    int variable2;             //友好变量
    private int variable3;     //私有变量
    public int method1(){      //公共方法
        return 1;
    }
    int method2(){             //友好方法
        return 2;
    }
    private int method3(){     //私有方法
        return 3;
    }
}
public class Task3_14 {
    public static void main(String[] args) {
        AccessControl a1 = new AccessControl();
        //问题1: 语句"a1.variable2 = 2;"是否合法? 说明理由
        //问题2: 语句"a1.variable3 = 3;"是否合法? 说明理由
        //问题3: 语句"AccessControl.variable1 = 3;"是否合法? 说明理由
        //问题4: 语句"int b = a1.method2;"是否合法? 说明理由
        //问题5: 语句"int c = a1.method3;"是否合法? 说明理由
        task3_14.package1.AccessControl a2 = new task4_12.package1.AccessControl();
        //问题6: 语句"a2.variable1 = 2;"是否合法? 说明理由
        //问题7: 语句"a2.variable2 = 3;"是否合法? 说明理由
        //问题8: 语句"int b = a2.method1;"是否合法? 说明理由
        //问题9: 语句"int c = a2.method2;"是否合法? 说明理由
    }
}
```

图 3-34　Task3_14.java 文件代码

回答 Task3_14 类中 main 方法中的 9 个问题,答案直接写在图 3-34 实验作业提交界面中。

(十二) 类的访问权限

实验内容：通过实验理解类的访问权限

使用 Eclipse 创建 Java 项目"task3_15",在该项目中创建一个名为"PublicClass"的 Java 类,在"New Java Class"对话框中指定 AccessControl 类的包名为"task3_15.package1",效果如图 3-35 所示。PublicClass 类的代码如图 3-36 所示(还包括名为 FriendClass 的友好类)。

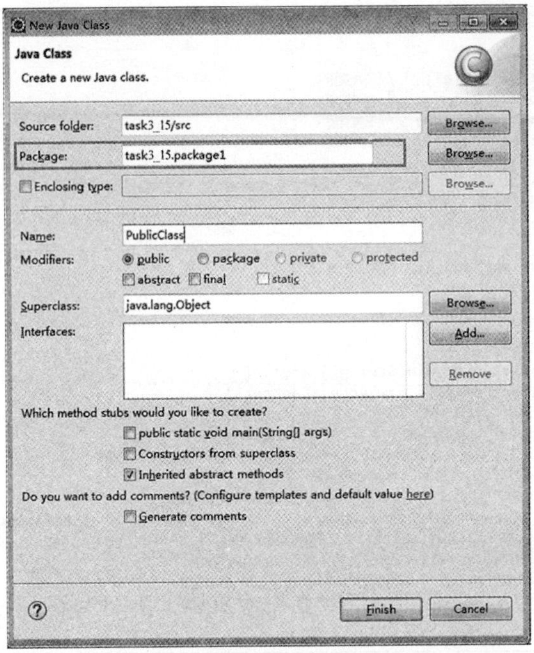

图 3-35　指定 AccessControl 类的包名为 task3_15.package1 的界面

```
package task3_15.package1;

public class PublicClass {          //公共类
    public int variable1;
    int variable2;
    private int variable3;
}

class FriendClass{                  //友好类
    public int variable1;           //公共变量
    int variable2;                  //友好变量
    private int variable3;          //私有变量
}
```

图 3-36　Task3_15.package1 文件代码

在项目 task3_15 中创建一个名为"Task3_15"的 Java 类,在"New Java Class"对话框中指定 Task3_15 类的包名为"task3_15",效果如图 3-37 所示。Task3_15.java 文件中的代码如图 3-38 所示(还包括名为 FriendClass2 的友好类)。

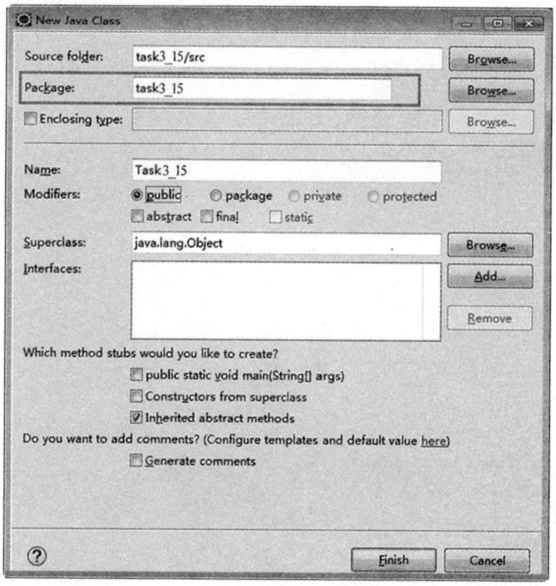

图 3-37　在"New Java Class"对话框中指定 Task3_15 类的包名界面

```
package task3_15;
class FriendClass2{        //友好类
    public int variable1;
    int variable2;
    private int variable3;
}
public class Task3_15 {
    public static void main(String[] args) {
        //问题1：语句"task3_15.PublicClass a;"是否合法?说明理由
        //问题2：语句"task3_15.FriendClass b;"是否合法?说明理由
        //问题3：语句"FriendClass2 c;"是否合法?说明理由
    }
}
```

图 3-38　Task3_15.java 文件代码

回答 Task3_15 类中 main 方法中的 3 个问题,答案直接写在图 3-38 实验作业提交界面中。

(十三) 使用修饰符

有时需要公开一些变量和方法,有时需要禁止其他对象使用一些变量和方法,这时可以使用修饰符来实现这个目的。常用的修饰符有 public,private,protected,package,static,final,transient,volatile 等。

1. 程序功能

通过两个类 StaticDemo、Task3_16 说明静态变量/方法与实例变量/方法的区别。

2. 编写类文件 Task3_16.java

程序源代码如下。

```java
class StaticDemo {
    static int x;
    int y;
    public static int getX() {
        return x;
    }
    public static void setX(int newX) {
        x = newX;
    }
    public int getY() {
        return y;
    }
    public void setY(int newY) {
        y = newY;
    }
}
public classTask3_16{
    public static void main(String[] args) {
        System.out.println("静态变量 x="+StaticDemo.getX());
        System.out.println("实例变量 y="+StaticDemo.getY()); // 非法,编译时将出错
        StaticDemo a= new StaticDemo();
        StaticDemo b= new StaticDemo();
        a.setX(1);
        a.setY(2);
        b.setX(3);
        b.setY(4);
        System.out.println("静态变量 a.x="+a.getX());
        System.out.println("实例变量 a.y="+a.getY());
```

```
        System.out.println("静态变量 b.x="+b.getX());
        System.out.println("实例变量 b.y="+b.getY());
    }
}
```

3. 编译并运行

对上面的源程序进行编译,排错并运行。

四、思考题

(1) 说明使用变量之前是不是都要先声明变量或使用变量之前是不是都要先赋值,为什么?

(2) 说明什么是构造方法。

(3) 说明程序中有多个类时如何确定源程序文件的名称。

实验四 数组、向量和字符串

一、实验目的

- 掌握 Java 中的数组定义、引用。
- 掌握向量的基本概念和应用技术。
- 掌握使用字符串 String 类处理字符串的方法。
- 掌握使用字符串 StringBuffer 类处理字符串的方法。

二、实验要求

- 编写一个使用 Java 数组的程序。
- 掌握向量类的使用方法。
- 掌握字符串类的使用方法。

三、实验内容

(一) 使用数组

1. 建立使用数组的程序

本程序建立了一个长度为 5 的 1 维数组,一个长度为 12 的 2 维数组。

(1) 源代码如下。

```
public class KY4_1 {
```

```java
public static void main(String args[]) {
    int a[]=new int[5];
    int arr1[][]=new int[3][4];
    a[0]=10;
    a[1]=10+a[0];
    a[2]=30;
    a[3]=40;
    a[4]= a[1]+ a[2];
    arr1[0][0]=0; arr1[0][1]=1; arr1[0][2]=2;
    arr1[1][0]=3; arr1[1][1]=4; arr1[1][2]=5;
    arr1[2][0]=6; arr1[2][1]=7; arr1[2][2]=8;
    System.out.println("a["+0+"] = "+a[0]);
    System.out.println("a["+1+"] = "+a[1]);
    System.out.println("a["+2+"] = "+a[2]);
    System.out.println("a["+3+"] = "+a[3]);
    System.out.println("a["+4+"] = "+a[4]);
    System.out.println("arr1("+0+","+0+") = "+arr1[0][0]);
    System.out.println("arr1("+0+","+1+") = "+arr1[0][1]);
    System.out.println("arr1("+0+","+2+") = "+arr1[0][2]);
    System.out.println("arr1("+1+","+0+") = "+arr1[1][0]);
    System.out.println("arr1("+1+","+1+") = "+arr1[1][1]);
    System.out.println("arr1("+1+","+2+") = "+arr1[1][2]);
  }
}
```

（2）编译并运行程序。

2. 编程实现 Fibonacci 数列

（1）Fibonacci 数列的可作如下定义。

$F_1=1, F_2=1, \cdots, F_n=F_{n-1}+F_{n-2}$　　　　$(n \geqslant 3)$

（2）关键代码如下。

```
f[0]=f[1]=1;
for(i=2;i<10;i++)
```

$$f[i]=f[i-1]+f[i-2];$$

3. 编程采用冒泡法实现对数组元素由小到大排序

（1）冒泡法排序是对相邻的两个元素进行比较，并把小的元素交换到前面。

（2）关键代码如下。

```
for(i=0;i<intArray.length-1;i++)
    for(j=i+1;j<intArray.length;j++)
        if(intArray[i]>intArray[j]){
            t=intArray[i];intArray[i]=intArray[j];intArray[j]=t;
        }
```

（二）使用向量类

大多数编程语言中的数组是固定长度的，即数组一经建立就不能在使用过程中改变其长度。Java 引入 Vector 类来创建可以改变长度的变量。Vector 被设计成一个能不断增长的序列，它类似于可变长数组，但功能更加强大，因为任何类型的对象都可以放入 Vector 类的对象中。通过调用 Vector 封装的方法，可以随时添加或删除向量元素，以及增加或缩短向量序列的长度。

1. 创建应用程序

创建使用 Vector 向量类的应用程序

2. 程序功能

创建一个 Vector 对象 v，先通过键盘为 args[] 输入两个分量，然后赋值给对象 v，并通过直接方式为其分量赋值。

3. 编写 KY4_2.java 程序文件

源代码如下。

```java
import java.util.*;
public class KY4_2{
    public static void main(String args[]) {
        Vector v=new Vector(1,1);
        v.addElement(args[0]); //在向量尾部添加元素
        v.addElement(args[1]);
        v.addElement("3"); //在向量尾部添加元素
        v.insertElementAt("0",0);//在指定位置插入元素
```

```
        v.insertElementAt("aaa字符串元素",3);
        v.setElementAt("4",4);//替换指定位置的元素
        v.addElement("5");
        System.out.println("第4号元素为"+v.elementAt(4));
        Enumeration enum=v.elements();//枚举化对象,以便逐个取出元素
        StringBuffer buffer=new StringBuffer();//字符串缓冲区
        while(enum.hasMoreElements())
        buffer.append(enum.nextElement()).append(",");
        buffer.deleteCharAt(buffer.length()-1);
        System.out.println("向量v的所有元素:"+buffer.toString()+"\n");
        System.out.println("向量v的元素个数="+v.size()+",v的长度为"+v.capacity()+"\n");
        v.removeAllElements();
        System.out.println("删除后元素个数:"+v.size()+",向量v长度:"+v.capacity()+"\n");
    }
}
```

4. 编译并运行程序

编译并运行以上程序。

(三) 使用字符串与字符串类

1. 实验内容1：编写 KY4_3.java 程序文件

源代码如下。

```
import java.applet.Applet;
import java.awt.Graphics;
public class KY4_3 extends Applet {
  public void paint(Graphics g) {
      String str="这是一个字符串——This is a test string";
      g.drawString("正常字符串:"+str,30,30);
      g.drawString("翻转字符串:"+reverse(str),30,60);
  }
  public String reverse(String s) {
      int len=s.length();
      StringBuffer buffer=new StringBuffer(len);
```

```
        for (int i=len-1; i>=0; i--)
        buffer.append(s.charAt(i));
        return buffer.toString();
    }
}
```

(1) 编译 KY4_3.java 程序文件。

(2) 编写显示 KY4_3.class 的页面文件 KY4_3.html,源代码如下。

```
<html>
<applet codebase=e:/java/程序 code=KY4_3.class width=400 height=120>
    </applet>
</html>
```

(3) 在浏览器打开 KY4_3.html 文件。

2. 实验内容 2:通过实验掌握 String 类常用方法的使用

使用 Eclipse 创建 Java 项目"task4_3A",在该项目中创建一个名为"Task4_3A"的 Java 主类。

在 Task4_3A 类的 main 方法中,实现如下功能:程序运行后,提示"请输入一个字符串:",用户输入一个字符串后按 Enter 键,程序判断该字符串是否是回文(左右对称的字符串,如 123454321),并显示判断结果。然后再次提示"请输入一个字符串:",并再次判断新输入的字符串是否是回文,直到用户输入的字符串为"exit"时,退出程序。程序流程如图 4-1 所示。程序运行效果参见 Task4_3A.class(下载后保存到 D 盘根目录,并在 MS-DOS 窗口输入命令执行"java Task4_3A")。

3. 实验内容 3:通过实验掌握 String 类对象与字符数组的相互转换

使用 Eclipse 创建 Java 项目"task4_3B",在该项目中创建一个名为"Task4_3B"的 Java 主类,"Task4_3B.java"文件中的代码如图 4-2 所示(其中包含字符串加密解密

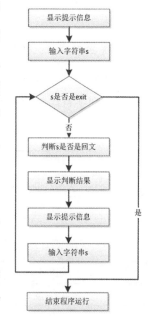

图 4-1　判断回文程序流程

类 MyEncryption)。

请将图 4-2 所示代码中涂黑的部分补充完整,使程序能够正常运行。

```
//字符串加密解密类
class MyEncryption{
    private String password;              //成员变量,用于保存密码
    public void setPassword(String pwd){  //成员方法,用于设置密码
        password = pwd;
    }
    public String EncodeStr(String s){    //成员方法,用于对字符串s进行异或加密
        char[] c = _____;              //将文字串s转换为字符数组
        char[] p = _____;              //将密码password转换为字符数组
        for(int i=0,j=0;i<c.length;i++,j++){
            if(j>=p.length) j=0;          //循环使用密码
            c[i]= (char)(c[i] ^ p[j]);    //进行异或运算
        }
        _____;                         //返回加密后的字符串
    }
    public String DecodeStr(String s){    //成员方法,用于对字符串s进行异或解密
        return EncodeStr(s);              //异或解密和异或加解密的算法是相同的
    }
    public static void main(String[] args) {
        MyEncryption e = new MyEncryption();
        String s = "My name is XXX";
        e.setPassword("wbgsn");
        String encodedStr = e.EncodeStr(s);
        System.out.println("原文内容如下: "+ s);
        System.out.print("加密后的密文: ");
        System.out.println(encodedStr);
        String decodedStr = e.DecodeStr(encodedStr);
        System.out.print("解密后的原文: ");
        System.out.println(decodedStr);
    }
}
```

图 4-2 Task4_3.java 文件代码

4. 实验内容 4:理解正则表达式的功能

已知有正则表达式"(0|[1-9][0-9]*)([.][0-9]*[1-9])?",请在表 4-1 中写出下列字符串与该正则表达式是否匹配。

表 4-1 字符串与正则表达式匹配分析

序号	字符串	填"匹配"或"不匹配"
1	01.23	
2	1.230	
3	0.103	
4	101.4	
5	1234.	
6	1.2.3	
7	.1234	
8	123.0	

通过实验验证上面的答案,使用 Eclipse 创建 Java 项目"task4_3C",在该项目中创建一个名为"Task4_3C"的 Java 主类,"Task4_3C.java"文件中的代码如图 4-3 所示。

```java
public static void main(String[] args) {
    String regex = "(0|[1-9][0-9]*)([.][0-9]*[1-9])?";   //正则表达式
    Scanner reader = new Scanner(System.in);
    System.out.println("请输入一个字符串:");
    String s = reader.nextLine();
    while(!s.equals("exit")){
        if(s.matches(regex)){
            System.out.println("字符串""+s+"" 与正则表达式""+regex+"" 匹配");
        }
        else{
            System.out.println("字符串""+s+"" 与正则表达式""+regex+"" 不匹配");
        }
        System.out.println("请输入一个字符串:");
        s = reader.nextLine();
    }
    reader.close();
    System.out.println("程序已经退出!");
}
```

图 4-3 Task4_3C.java 文件代码

思考上述实验内容的答案,格式如下。

字符串 1:"匹配"或"不匹配"。

字符串 2:"匹配"或"不匹配"。

字符串 3:"匹配"或"不匹配"。

……

字符串 8:"匹配"或"不匹配"。

5. 实验内容 5:通过实验掌握正则表达式的使用

使用 Eclipse 创建 Java 项目"task4_3D",在该项目中创建一个名为"Task4_3D"的 Java 主类,"Task4_3D.java"文件中的代码如图 4-4 所示。

```java
public static void main(String[] args) {
    String regex = ▓▓▓▓▓▓ ;  //正则表达式,表示连续的若干非英文字母
    String s = "My123name 34^5is    Yuan Liyong";//需要处理的字符串
    String[] ss = ▓▓▓▓▓▓ ;//使用正则表达式regex把字符串s分解为一个字符串数组
    //请补充代码,实现输出字符串数组的元素个数
    //请补充代码,输出字符串数组的内容
}
```

图 4-4 Task4_3D.java 文件代码

将图4-5所示代码中的涂黑部分补充完整,使程序能够输出如图4-5所示内容。

图4-5 Task4_3D.java文件运行结果

根据下列要求写出相应的正则表达式(选做)。

要求1:写出用于匹配1个字符串是否是3位整数(数值在100到999之间,可以是负数)的正则表达式。

要求2:写出用于匹配1个字符串是否是1个十六进制数(十六进制数以0x开头,后面是0到9及A到F(小写也可以),例如0x0A、0xa2等)的正则表达式。

要求3:写出用于匹配1个字符串是否是合法的Java标识符的正则表达式。

要求4:写出用于匹配1个电子邮箱地址是否是浙江师范大学电子邮箱地址(********@zjnu.cn 或 ********@zjnu.edu.cn,其中********可以是英文字母、数字和下划线的组合,最长不超过16个字符)的正则表达式。

思考上述实验内容的答案,格式如下。

要求1:正则表达式。

要求2:正则表达式。

要求3:正则表达式。

要求4:正则表达式。

6. 实验内容6:掌握StringBuffer类的使用

已知有如图4-6所示代码,请回答图中的7个问题。

数组、向量和字符串 实验四

```
StringBuffer buffer = new StringBuffer(20);
buffer.append("yuan");
buffer.append("浙江师范大学");
System.out.println(buffer);                    //问题1：输出内容是什么？
System.out.println(buffer.length());           //问题2：输出内容是什么？
System.out.println(buffer.capacity());         //问题3：输出内容是什么？
buffer.replace(4, 10, ".zjnu.cn");
System.out.println(buffer);                    //问题4：输出内容是什么？
System.out.println(buffer.charAt(4));          //问题5：输出内容是什么？
buffer.setCharAt(4, '@');                      //设置
System.out.println(buffer);                    //问题6：输出内容是什么？
buffer.insert(9, ".edu");
System.out.println(buffer);                    //问题7：输出内容是什么？
```

图 4-6　Task4_3E.java 文件的代码

通过实验验证上面的答案，使用 Eclipse 创建 Java 项目"task4_3E"，在该项目中创建一个名为"Task4_3E"的 Java 主类，"Task4_3E.java"文件中的 main 方法的代码如图 4-6 所示。

思考上述实验内容的答案，格式如下：

问题 1：输出内容。

问题 2：输出内容。

问题 3：输出内容。

……

问题 7：输出内容。

四、思考题

（1）使用数组存储一个英文句子："Java is an objected oriented programming language"，显示该句子，并算出每个单词的平均字母数。

（2）有关于字符串的方法很多，请试着分析方法 toUpperCase 是如何实现的？

实验五 继承性和多态性

一、实验目的

- 掌握继承的实现方法,理解继承的概念。
- 理解成员变量和方法的可继承性。
- 掌握多态性的基本概念和应用技术。
- 成员变量的隐藏和方法的重写。

二、实验要求

- 完成各实验内容的程序。
- 掌握类的继承的使用方法。
- 掌握类的多态性的使用方法。

三、实验内容

(一) 类的继承

1. 实验内容1:类继承的实现方法

使用 Eclipse 创建名为"task5_1"的 Java 项目,在该项目中创建一个名为"Person"的 Java 类,该类实现的图 5-1 所示 Person 类 UML 类图的功能。在"task5_1"项目中再创建一个名为"Student"的 Java 类,Student 类继承于 Person

类,同时实现如图 5-1 所示 Student 类 UML 类图的功能。其中,Person 类的 introduce 方法和 Student 类的 tellStudentNo 方法的输出效果如图 5-2 所示。

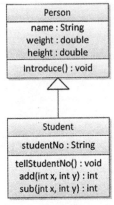

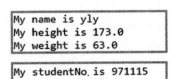

图 5-1　Person 类和 Student 类的 UML 类图

图 5-2　introduce 方法和 tell Student No. 方法输出效果

2. 实验内容 2:理解和验证继承的概念

在"task5_1"项目中再创建一个名为"Task5_1"的 Java 主类,Task5_1 类的 main 方法的代码如图 5-3 所示。回答 Task5_1 类 main 方法中的 4 个问题,答案直接写在实验作业提交界面中。

```
public static void main(String args[]) {
    Person p = new Person();
    Student s = new Student();
    int c;
    //问题1:语句"s.height = 173;"是否合法?并说明理由
    //问题2:语句"s.introduce();"是否合法?并说明理由
    //问题3:语句"c = s.sub(23,21);"是否合法?并说明理由
    //问题4:语句"c = p.add(12,23);"是否合法?并说明理由
}
```

图 5-3　Task5_1 类的 main 方法代码

(二) 成员变量和方法的可继承性

1. 实验内容 1:通过实验验证父类成员变量和方法的可继承性(在子类中的可见性)

使用 Eclipse 创建 Java 项目"task5_2",在该项目中创建一个名为"Parent"的 Java 类,在"New Java Class"对话框中指定 Parent 类的包名为"task5_2.package1",效果如图 5-4 所示。Parent 类的代码如图 5-5 所示。

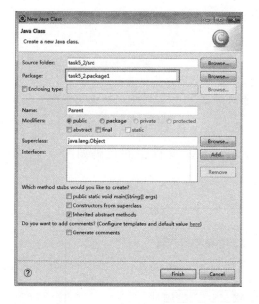

图 5－4　创建包名为 task5_2.package1 的 Parent 类

```
package task5_2.package1;
public class Parent {
    public int var1;
    protected int var2;
    int var3;
    private int var4;
    protected void method1(){
        //代码流程图
    }
    void method2(){
        //代码流程图
    }
}
```

图 5－5　task5_2.package1 代码

在项目 task5_2 中再创建一个名为“Parent”的 Java 类，在“New Java Class”对话框中指定该 Parent 类的包名为“task5_2”，效果如图 5－6 所示。Parent.java 文件中的代码如图 5－7 所示（包含了 Child1 类、Child2 类的代码）。

图 5－6　创建包名为 task5_2 的 Parent 类

```
package task5_2;
public class Parent {
    protected int var1;
    int var2;
    private int var3;
    void method1(){          //代码流程图
    }
    private void method2(){ //代码流程图
    }
}
//Child1类继承自同一个包中的Parent类（即上面的Parent类）
class Child1 extends Parent{
    void testInheritable(){
        //问题1：语句“var1 = 1;”是否合法？
        //问题2：语句“var2 = 1;”是否合法？
        //问题3：语句“var3 = 1;”是否合法？
        //问题4：语句“method1();”是否合法？
        //问题5：语句“method2();”是否合法？
    }
}
//Child1类继承自不同包中的Parent类
class Child2 extends task5_2.package1.Parent{
    void testInheritable(){
        //问题6：语句“var1 = 1;”是否合法？
        //问题7：语句“var2 = 1;”是否合法？
        //问题8：语句“var3 = 1;”是否合法？
        //问题9：语句“method1();”是否合法？
        //问题10：语句“method2();”是否合法？
    }
}
```

图 5－7　Parent.java 文件代码

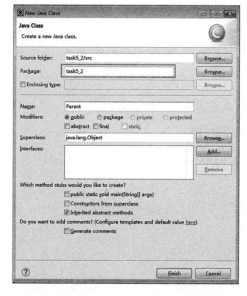

回答 Child1 类和 Child2 类中代码部分的 10 个问题,答案直接写在图 5-7 实验作业提交界面中。

（三）成员变量的隐藏和方法的重写

1. 实验内容 1：理解和掌握成员变量的隐藏

使用 Eclipse 创建 Java 项目"task5_3",在该项目中创建一个名为"HiddenMember"的 Java 类。HiddenMember.java 文件中的代码如图 5-8 所示（其中包含了 Parent 类、Child 类的代码）。

```
package task5_3;
class Parent{
    double member = 3.14;
    double getHiddenMember(){    //父类方法只能访问父类的变量
        return member;
    }
}
class Child extends Parent{
    int member;    //子类声明的变量会隐藏（屏蔽）父类的同名变量
    int getMember(){
        return member;    //返回的是子类定义的成员变量
    }
}
public class HiddenMember {
    public static void main(String[] args) {
        Child c = new Child();
        c.member = 10;
        System.out.println(c.getMember());        //问题1：该语句输出的值是什么？
        System.out.println(c.getHiddenMember());  //问题2：该语句输出的值是什么？
        //问题3：语句"c.member = 10.0;"是否合法？说明理由
    }
}
```

图 5-8 HiddenMember.java 文件代码

回答 HiddenMember 类 main 方法中的 3 个问题,答案直接写在图 5-8 实验作业提交界面中。

2. 实验内容 2：理解和掌握方法的重写（或覆盖）

在"task5_3"项目中创建一个名为"RewritedMethod"的 Java 类。RewritedMethod.java 文件中的代码如图 5-9 所示（其中包含了 University 类、ImportantUniversity 类的代码）。

回答 RewritedMethod 类 main 方法中的 2 个问题,答案直接写在图 5-9 实验作业提交界面中。

继承性和多态性 实验五

```
package task5_3;
class University{
    public void enterRule(double math,double english,double chinse){
        double total = math + english+chinse;
        if(total>=200)
            System.out.println("总分为"+total+"分,达到大学的最低分数线,录取!");
        else
            System.out.println("总分为"+total+"分,没有达到大学的最低分数线,拒绝!");
    }
}
class ImportantUniversity extends University{
    //ImportantUniversity类重写了父类的enterRule方法
    public void enterRule(double math,double english,double chinse){
        double total = math + english+chinse;
        if(total>=250)
            System.out.println("总分为"+total+"分,达到重点大学的最低分数线,录取!");
        else
            System.out.println("总分为"+total+"分,没有达到重点大学的最低分数线,拒绝!");
    }
}
public class RewritedMethod {
    public static void main(String[] args) {
        University univ = new University();
        univ.enterRule(60,80,70);            //问题1:调用此方法后,输出的内容是什么?
        ImportantUniversity importantUniv = new ImportantUniversity();
        importantUniv.enterRule(60, 80, 70); //问题2:调用此方法后,输出的内容是什么?
    }
}
```

图 5-9　RewritedMethod.java 文件代码

（四）使用 super 关键字

1. 实验内容 1：使用 super 关键字访问被子类隐藏的父类成员变量和被子类重写的父类方法

使用 Eclipse 创建 Java 项目"task5_4"，在该项目中创建一个名为"Task5_4"的 Java 类。Task5_4.java 文件中的代码如图 5-10 所示（其中包含了 Bank 类、ConstructionBank 类的代码）。

```
class Bank{
    int savedMoney;      //表示存款金额
    int year;            //表示存的年数（整数）
    double computeInterest(){   //按整年计算利息
        return savedMoney * year * 0.04;
    }
}
class ConstructionBank extends Bank{
    double year; //year=3.150表示存了3年123天,子类的double类型变量year隐藏了父类的同名变量
    double computeInterest(){   //ConstructionBank类重写了父类Bank类的同名方法
        //填空1:此处填写一句代码,year的整数值"(int)year"赋值给父类的year变量
        double total;
        //填空2:此处填写一句代码,调用父类方法computeInterest()并将计算结果赋值给变量total
        int remainder = (int)((year-(int)year)*1000);    //计算不足一年的天数
        total += remainder * 0.0001 * savedMoney;
        System.out.println(savedMoney+"存了"+super.year+"年"+remainder+"天");
        return total;
    }
}
public class Task5_4{
    public static void main(String args[]){
        ConstructionBank account = new ConstructionBank();
        account.savedMoney = 10000;
        account.year = 3.150;
        System.out.println(account.computeInterest());
    }
}
```

图 5-10　Task5_4.java 文件代码

在 ConstructionBank 类的 computeInterest 方法中的"填空 1"和"填空 2"处填入正确的代码。

2. 实验内容 2：使用 super 关键字调用父类的构造方法

在"task5_4"项目中创建一个名为"Constructor"的 Java 类。Constructor.java 文件中的代码如图 5-11 所示（其中包含了 Person 类、Student 类的代码）。

```java
class Person{
    private String cardID;        //表示身份证号
    String name;                  //表示姓名
    String getCardID(){
        return cardID;
    }
    public Person(String id) {    //Person类的构造方法
        cardID = id;
    }
}
class Student extends Person{
    private String studentID;
    String getStudentID(){
        return studentID;
    }
    Student(String cID,String sID){   //Student类的构造方法
        //填空1:在此处填写一条语句，调用父类的构造函数
        studentID = sID;
    }
}
public class Constructor {
    public static void main(String[] args) {
        Student s1 = new Student("330623780", "971115");
        s1.name = "yly";
        System.out.println(s1.name+"'s cardID is "+s1.getCardID());
        System.out.println(s1.name+"'s studentID is "+s1.getStudentID());
    }
}
```

图 5-11 Constructor.java 文件代码

在 Student 类的构造方法中的"填空 1"处填入正确的代码。

（五）使用 final 关键字

实验内容：验证 final 关键字修饰类、成员变量和成员方法的功能

使用 Eclipse 创建 Java 项目"task5_5"，在该项目中创建一个名为"Task5_5"的 Java 类。Task5_5.java 文件中的代码如图 5-12 所示（其中包含了 FinalClass 类、Circle 类和 AbsoluteCircle 类的代码）。

```java
final class FinalClass{
    int var1;
    int method1(){
        return var1;
    }
}
//问题1：是否可以编写一个继承自FinalClass的类
class Circle{
    final double PI=3.14;        //final变量
    double radius;
    final double getArea(){      //final方法
        //问题2：语句"PI=3.14159265;"是否合法？
        return radius * radius * PI;
    }
}
class AbsoluteCircle extends Circle{
    int centerX;
    int centerY;
    boolean isInCircle(int x,int y){   //判断坐标x,y是否在圆内
        if((x-centerX)*(x-centerX)+(y-centerY)*(y-centerY)<radius*radius)
            return true;
        else
            return false;
    }
    //问题3：是否可以对父类的getArea方法进行重写？
}
public class Task5_5 {
    public static void main(String args[]){
        AbsoluteCircle aCircle = new AbsoluteCircle();
        aCircle.centerX = 10;
        aCircle.centerY = 10;
        aCircle.radius = 15;
        //问题4：语句"double s = aCircle.getArea();"是否合法？
        if(aCircle.isInCircle(0, 0))
            System.out.println("坐标(0,0)在圆内");
        else
            System.out.println("坐标(0,0)不在圆内");
    }
}
```

图 5-12　Task5_5.java 文件代码

回答代码中的 4 个问题，答案直接写在图 5-12 实验作业界面空白处。

（六）对象的上转型对象

实验内容：验证对象的上转型对象的功能特点

使用 Eclipse 创建 Java 项目"task5_6"，在该项目中创建一个名为"Task5_6"的 Java 类。Task5_6.java 文件中的代码如图 5-13 所示（其中包含了 Person 类和 Student 类的代码）。

```
package task5_6;
class Person{
    private String cardID;
    String name="无名氏";
    public Person(String id) {//构造方法，参数id用于初始化成员变量cardID
        cardID = id;
    }
    String getCardID(){
        return cardID;
    }
    void introduceSelf(){
        System.out.println("My cardID is "+cardID);
        System.out.println("My name is "+name);
        System.out.println("Person.introduceSelf()");
    }
}
class Student extends Person{
    private String studentID;
    String schoolName;
    //构造方法，参数cID用来初始化父类的成员变量cardID,参数sID用来初始化成员变量studentID
    public Student(String cID,String sID) {
        super(cID);
        studentID = sID;
    }
    void introduceSelf(){              //重写了从父类继承的同名方法
        super.introduceSelf();         //调用父类的方法
        System.out.println("My stduentID is "+studentID);
        System.out.println("Student.introduceSelf()");
    }
    String getStudentID(){
        return studentID;
    }
}
public class Task5_6 {
    public static void main(String[] args) {
        //对象p就是Student类对象的上转型对象
        Person p = new Student("33062378", "971115");
        //问题1：语句"Student s = (Student)(new Person("33062378"));"是否合法？说明理由
        //问题2：语句"p.name = "yly";"是否合法？说明理由
        //问题3：语句"System.out.println(s.getCardID());"是否合法？说明理由
        //问题4：语句"System.out.println(s.getStudentID());"是否合法？说明理由
        //问题5：语句"p.SchoolName = "ZJNU";"是否合法？说明理由
        //问题6：执行语句"p.introduceSelf();"后，输出的内容中包含什么？（从下面2个选项中选择1项）
        //选项（A）"Person.introduceSelf()"
        //选项（B）"Student.introduceSelf()"
    }
}
```

图 5-13　Task5_6.java 文件代码

回答代码中的 6 个问题，答案直接写在图 5-13 实验作业空白处。

（七）多态

实验内容：通过实验理解多态

使用 Eclipse 创建 Java 项目"task5_7"，在该项目中创建一个名为"Task5_7"的 Java 类。Task5_7.java 文件中的代码如图 5-14 所示（其中包含了 Human 类、Chinese 类和 American 类的代码）。

```
package task5_7;
class Human{
    double height;
    double weight;
    void introduce(){
        System.out.println("……¥#*（*—*&%……¥#");
    }
}
class Chinese extends Human{
    void introduce() {
        System.out.println("我是中国人");
        System.out.println("我的身高是"+height+"厘米");
        System.out.println("我的身体是"+weight+"公斤");
    }
}
class American extends Human{
    void introduce() {
        System.out.println("I am American");
        System.out.println("My height is "+height+"cm");
        System.out.println("My weight is "+weight+"kg");
    }
}
public class Task5_7 {
    public static void main(String[] args) {
        Human human = new Chinese();
        human.height = 173;
        human.weight = 63;
        human.introduce();      //问题1：这条语句输出的内容是什么？
        human= new American();
        human.height = 173;
        human.weight = 63;
        human.introduce();      //问题2：这条语句输出的内容是什么？
    }
}
```

图 5-14　Task5_7.java 文件代码

写出项目 Task5_7 运行后输出的内容，答案直接写在图 5-14 界面空白处。

（八）抽象类

实验内容：通过实验理解和掌握抽象类的使用

使用 Eclipse 创建 Java 项目"task5_8"，在该项目中创建一个名为"Task5_8"的 Java 类。Task5_8.java 文件中的代码如图 5-15 所示（其中包含了 EspecialCar 抽象类，其子类 PoliceCar 类、AmbulanceCar 类、FireCar 类，以及通用模块 Simulator 类的代码）。

```
package task5_8;
abstract class EspecialCar{        //声明特种车辆抽象类
    int seatCount;                  //实例变量
    int getSeatCount(){             //实例方法
        return seatCount;
    }
    abstract void cautionSound();   //抽象方法,没有方法体
}
class PoliceCar extends EspecialCar{      //警车类
    void cautionSound(){            //警车警报声
        System.out.println("警车的警笛声: zhua..zhua..zhua..");
    }
}
class AmbulanceCar extends EspecialCar{   //救护车类
    void cautionSound(){            //
        System.out.println("救护车的救护声: qiu..qiu..qiu..");
    }
}
class FireCar extends EspecialCar{        //消防车类
    void cautionSound(){            //
        System.out.println("消防车的救火声: huo..huo..huo..");
    }
}
class Simulator{       //通用化的程序模块
    void simulate(EspecialCar eCar){      //将抽象类对象作为参数
        eCar.cautionSound();
    }
}
public class Task5_8 {
    public static void main(String[] args) {
        //问题1: 语句"EspecialCar car = EspecialCar()"是否合法?说明理由
        Simulator simulator = new Simulator();
        System.out.println("当前模拟的是警车的声音");
        simulator.simulate(new PoliceCar());
        System.out.println("当前模拟的是救护车的声音");
        simulator.simulate(new AmbulanceCar());
        System.out.println("当前模拟的是消防车的声音");
        //问题2: 填写一条语句,用来模拟消防车的声音
    }
}
```

图 5-15　Task5_8.java 文件代码

回答图 5-15 所示代码中的问题,答案直接写在实验作业界面空白处。

（九）面向抽象编程 A

实验内容:通过实验理解面向抽象编程的基本思想(在关联关系中使用抽象类),编程实现图 5-16 所示的面向抽象的 UML 图软件框架

使用 Eclipse 创建 Java 项目"task5_9",在该项目中创建一个名为"Task5_9"的 Java 类。Task5_9.java 文件中的代码如图 5-17 所示(其中包含了 Geometry 抽象类,其子类 Circle 类和 Rectangle 类,以及 Pillar 类的代码)。

继承性和多态性 实验五

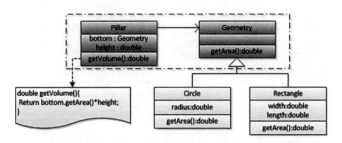

图 5-16　Task5_9.java 面向抽象的 UML 图软件框架

```
package task5_9;
abstract class Geometry{          //抽象类Geometry
    abstract double getArea();    //计算面积的抽象方法
}
class Pillar{                     //Pillar类能够表示各种柱体
    Geometry bottom;              //将抽象类对象作为其成员变量
    double height;
    double getVolume(){
        return bottom.getArea()*height;
    }
}
class Circle extends Geometry{
    double r;
    double getArea(){
        return r*r*3.14;
    }
}
class Rectangle extends Geometry{
    double length;
    double width;
    double getArea(){
        return width*length;
    }
}
public class Task5_9 {            //用户程序
    public static void main(String[] args) {
        Circle c1 = new Circle();
        c1.r = 10;
        Pillar p1 = new Pillar();
        p1.height = 10;
        p1.bottom = c1;
        System.out.println(p1.getVolume());
        Rectangle r1 = new Rectangle();
        r1.width = 4;
        r1.length = 5;
        p1.bottom = r1;
        System.out.println(p1.getVolume());
    }
}
```

图 5-17　Task5_9.java 文件代码

（十）面向抽象编程 B

实验内容：通过实验理解面向抽象编程的基本思想（在依赖关系中使用抽象类）编程实现图 5-18 所示的面向抽象的 UML 图软件框架

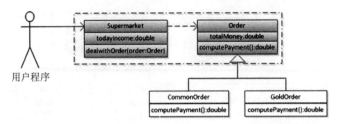

图 5-18 Task5_10 面向抽象的 UML 图软件框架

使用 Eclipse 创建 Java 项目"task5_10",在该项目中创建一个名为"Task5_10"的 Java 类。Task5_10.java 文件中的代码如图 5-19 所示(其中包含了 Order 抽象类,其子类 CommonOrder 类、GoldOrder 类,以及 Supermarket 类的代码)。

```
package task5_10;
abstract class Order{              //抽象类,购物单类
    double totalPrice;
    abstract double computePayment();
}
class Supermarket{                 //超市类
    static double todayIncome;
    static void dealWithOrder(Order order){   //抽象类作为方法的参数
        double payment = order.computePayment();
        System.out.println("需实付"+payment+"元");
        todayIncome += todayIncome;
    }
}
class CommonOrder extends Order{   //普通订单
    double computePayment(){
        System.out.println("普通客户,没有折扣!");
        return totalPrice;
    }
}
class GoldOrder extends Order{     //金卡订单
    double computePayment(){
        System.out.println("金卡客户,九折优惠!");
        return 0.9*totalPrice;
    }
}
//问题1:以抽象类Order为父类实现银卡订单,实际应付额为商品总价的9.5折
public class Task5_10 {
    public static void main(String[] args) {
        Supermarket.todayIncome = 0;
        //处理普通定单
        Order order = new CommonOrder();
        order.totalPrice = 190;
        Supermarket.dealWithOrder(order);
        //处理金卡定单
        order = new GoldOrder();
        order.totalPrice = 200;
        Supermarket.dealWithOrder(order);
        //问题2:处理银卡定单(商品价格为500元)
    }
}
```

图 5-19 Task5_10.java 文件代码

回答图 5-19 代码中的 2 个问题,答案直接写在实验作业提交界面中(银卡订单类名建议使用 Silverorder)。

四、思考题

(1) Java 中实现多态的机制是什么?

(2) Java 中继承的优缺点是什么?

实验六 包、接口与异常处理

一、实验目的

- 了解 Java 中包(package)、接口(interface)和异常处理(exception)的作用。
- 掌握包、接口、异常处理的设计方法。

二、实验要求

- 了解 Java 系统包的结构。
- 掌握创建自定义包的方法。
- 掌握使用系统接口的技术和创建自定义接口的方法。
- 理解系统异常处理的机制和创建自定义异常的方法。

三、实验内容

（一）了解并使用 Java 的系统包

包是类和接口的集合。利用包可以把常用的类或功能相似的类放在一个包中。Java 程序设计提供系统包，其中包含了大量的类，可以在编写 Java 程序时直接引用它们。为便于管理和使用它们，将这些类分为了不同的包。包又被称为类库或 API 包，所谓 API(Application Program Interface)即应用程序接口。API 包一方面提供丰富的类与方法供大家使用，如绘制图形、播放声音等，另一方面又负责和系统软硬件打交道,圆满实现用户程序的功能。所有 Java API 包都以"java."开头,以区别用户创建的包。

接口解决了 Java 不支持多重继承的问题,可以通过实现多个接口达到与多重继承相同的功能。

处理程序运行时的错误和设计程序同样重要,只有能够完善处理运行时出错的程序,软件系统才能长期稳定地运行,异常处理就是说明如何处理程序运行时出错的问题的。

(二)创建并使用自定义包

1. 自定义包的声明方式

　　＜package＞ ＜自定义包名＞

声明包语句必须添加在源程序的第一行,表示该程序文件声明的全部类都属于这个包。

2. 创建自定义包 Mypackage

在存放源程序的文件夹中建立一个子文件夹 Mypackage。例如,在"E:\java\程序"文件夹中创建一个与包同名的子文件夹 Mypackage(E:\java\程序\Mypackage),并将编译过的 class 文件放入该文件夹中。

注意:包名与文件夹名大小写要一致。

再添加环境变量 classpath 的路径,如下所示。

　　E:\j2sdk1.4.2_01\lib;E:\java\程序

3. 在包中创建类

(1) YMD.java 程序功能主要体现在:在源程序中,首先声明使用的包名 Mypackage,然后创建 YMD 类,该类具有计算今年的年份,并可以输出一个带有年月日的字符串的功能。

(2) 编写 YMD.java 文件,源代码如下。

```
package Mypackage;                //声明存放类的包
import java.util.*;               //引用 java.util 包
public class KY6_1_YMD {
    private int year,month,day;
    public static void main(String[] arg3){}
```

```java
    public KY4_1_YMD(int y,int m,int d) {
        year = y;
        month = (((m>=1) & (m<=12)) ? m : 1);
        day = (((d>=1) & (d<=31)) ? d : 1);
    }
    public KY4_1_YMD() {
        this(0,0,0);
    }
    public static int thisyear() {
        return Calendar.getInstance().get(Calendar.YEAR);//返回当年的年份
    }
    public int year() {
        return year;//返回年份
    }
    public String toString(){
        return year+"-"+month+"-"+day;//返回转化为字符串的年-月-日
    }
}
```

(3) 编译 KY6_1_YMD.java 文件,然后将 KY6_1_YMD.class 文件存放到 Mypackage 文件夹中。

3. 编写使用包 Mypackage 中 KY6_1_YMD 类的程序

(1) KY6_2.java 程序功能主要体现在:给定某人姓名与出生日期,计算该人年龄,并输出该人姓名、年龄、出生日期。程序使用了 KY6_1_YMD 的方法来计算年龄。

(2) 编写 KY6_2.java 程序文件,源代码如下。

```java
import Mypackage.KY4_1_YMD; //引用 Mypackage 包中的 KY4_1_YMD 类
public class KY4_2
{
    private String name;
    private KY4_1_YMD birth;
    public static void main(String args[])
```

```
        {
            KY6_2 a = new KY4_2("张驰",1990,1,11);
            a.output();
        }
        public KY6_2(String n1,KY4_1_YMD d1)
        {
            name = n1;
            birth = d1;
        }
        public KY6_2(String n1,int y,int m,int d)
        {
            this(n1,new KY6_1_YMD(y,m,d));//初始化变量与对象
        }
        public int age()                                //计算年龄
        {
            //返回当前年与出生年的差即年龄
            return KY6_1_YMD.thisyear() - birth.year();
        }
        public void output()
        {
            System.out.println("姓名："+name);
            System.out.println("出生日期："+birth.toString());
            System.out.println("今年年龄："+age());
        }
    }
```

（3）编译并运行程序。

（三）使用接口技术

1. 实现 MouseListener 和 MouseMotionListener 两个接口

（1）编写实现接口的程序文件 KY6_3.java，源代码如下。

```
import java.applet.Applet;
import java.awt.*;
```

```
        import java.awt.event.*;
public class KY4_3 extends Applet implements MouseListener,MouseMotionListener{
        int x1,y1,x2,y2;
        public void init(){
            addMouseListener(this);
            addMouseMotionListener(this);
        }
        public void paint(Graphics g){
            g.drawLine (x1,y1,x2,y2);
        }
        public void mousePressed(MouseEvent e){ //记录起点坐标
            x1=e.getX();
            y1=e.getY();
        }
        public void mouseClicked(MouseEvent e){}
        public void mouseEntered(MouseEvent e){}
        public void mouseExited(MouseEvent e){}
        public void mouseReleased(MouseEvent e){}
        public void mouseDragged(MouseEvent e){ //记录终点坐标
            x2=e.getX();
            y2=e.getY();
            repaint();
        }
        public void mouseMoved(MouseEvent e){}
}
```

(2) 编译 KY6_3.java 文件。

(3) 编写 KY6_3.html 文件,源代码如下。

```
<html>
<applet codebase=e：/java/程序 code=KY6_3.class width=320 height=180>
</applet>
</html>
```

(4) 在浏览器中打开 KY6_3.html 文件,在窗口中拖动鼠标可以随意画出一条线。

(四) 了解异常处理机制

1. 编写使用 try...catch 语句处理异常的程序文件 KY6_4.java 然后编译并运行程序

源代码如下。

```
public class KY6_4{
  public static void main(String[] arg3) {
    System.out.println("这是一个异常处理的例子\n");
    try {
      int i=10;
      i/=0;
    }
    catch (ArithmeticException e) {
      System.out.println("异常是:"+e.getMessage());
    }
    finally {
      System.out.println("finally 语句被执行");
    }
  }
}
```

注意:如果在 catch 语句中声明的异常类是 Exception,catch 语句也能正确地捕获,这是因为 Exception 是 ArithmeticException 的父类。如果不能确定会发生哪种情况的异常,那么最好指定 catch 的参数为 Exception,即说明异常的类型为 Exception。

2. 编写包含多个 catch 子句的 KY6_5.java 程序

源代码如下。

```
public class KY6_5{
  public static void main(String[] args) {
    try {
```

```
        int a=args.length;
        System.out.println("\na = "+a);
        int b=42/a;
        int c[]={1};
        c[42]=99;
    }
    catch (ArithmeticException e) {
        System.out.println("发生了被 0 除:"+e);
    }
    catch (ArrayIndexOutOfBoundsException e) {
        System.out.println("数组下标越界:"+e);
    }
  }
}
```

（五）接口的声明与使用

1. 实验内容 1：通过实验掌握接口的声明与使用

（1）使用 Eclipse 创建名为"task6_6"的 Java 项目。

（2）在"task6_6"项目中创建名为"Computable"的接口。在包浏览器（"Package Explorer"）中选择刚刚创建的项目"task6_6"，并右击，选择"New Java Interface"菜单项。系统将显示如图 6-1 所示的"Interface"对话框。

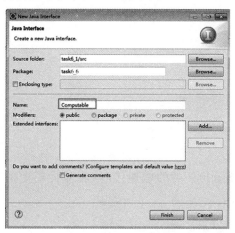

图 6-1　Interface 对话框

在图 6-1 中"Name"的位置输入接口名称"Computable",单击"Finish"按钮,系统完成创建接口文件"Helloworld.java"。然后,在 Computable 接口中定义两个抽象方法 add 和 sub,代码如图 6-2 所示。

```
public interface Computable {      //问题1:接口是否可以定义非抽象方法?
    public abstract int add(int a,int b);
    abstract int sub(int a,int b);   //问题2:该抽象方法是友好方法吗?
}
```

图 6-2 Helloworld.java 代码

(3)再创建一个名为"Student"的 Java 类来实现 Computable 接口。在包浏览器("Package Explorer")中选择项目"task6_7",并右击,选择"New Java Class"菜单项。系统将显示如图 6-2 所示的"Class"对话框。

在图 6-2 所示位置输入类名"Student",再通过单击"Add"按钮选择需要该类实现的接口(接口选择界面如图 6-3 所示),单击"Finish"按钮完成创建类文件"Student.java"。

图 6-2 "New Java Class"对话框

图 6-3 创建 Student.java 文件

补充完整生成的 Student 类的框架代码(删除代码中的"@Override",并实现 add 和 sub 方法),具体代码如图 6-4 所示。

```
public class Student implements Computable {
    String studentID;
    public int add(int a, int b) {        //实现接口中的方法
        return a + b;
    }
    public int sub(int a, int b) {        //实现接口中的方法
        return a - b;
    }
public static void main(String[] args) {
    //问题3：语句"Computable c = new Computable();"是否合法？说明理由
    Computable c = new Student();
    System.out.println(c.add(2, 38));
    //问题4：语句"System.out.println(c.studentID);"是否合法？说明理由
}
```

图 6-4　Student 类框架代码

（4）在"task6_7"项目中再创建一个名为"Task6_7"的 Java 主类，Task6_7 类的 main 方法的代码如图 6-4 所示。

2. 实验内容 2：理解接口及接口变量的使用

回答代码中的 4 个问题，答案直接写在图 6-2 和图 6-4 实验作业界面空白处。

（六）接口与多态

1. 实验内容：通过实验理解基于接口的多态

使用 Eclipse 创建 Java 项目"task6_8"，在该项目中创建一个名为"Task6_8"的 Java 类。Task6_8.java 文件中的代码如图 6-5 所示（其中包含了 Eatable 接口、Dog 类和 Cat 类的代码）。

```
package task6_8;
interface Eatable{                          //定义接口
    public abstract void sayFavoriteFood();
}
class Dog ▇▇▇▇ Eatable{   //问题1：被涂黑的部分填写什么关键字？
    public void sayFavoriteFood(){
        System.out.println("汪　汪汪,我最喜欢肉骨头");
    }
}
class Cat ▇▇▇▇ Eatable{   //问题1：被涂黑的部分填写什么关键字？
    public void sayFavoriteFood(){
        System.out.println("喵　喵喵,我最喜欢红烧鱼");
    }
}
public class Task6_8 {
    public static void main(String[] args) {
        Eatable e;
        Dog d = new Dog();
        Cat c = new Cat();
        e = d;
        e.sayFavoriteFood();   //问题2：请输出该语句的输出效果
        e = c;
        e.sayFavoriteFood();   //问题3：请输出该语句的输出效果
        //问题4：语句"d = c;"合法吗？
    }
}
```

图 6-5　Task6_8.java 文件代码

回答代码中的 4 个问题,答案直接写在图 6-5 实验作业界面空白处。

(七)面向接口编程 A

实验内容:通过实验理解面向接口编程的基本思想(把类的成员定义为接口变量)编程实现图 6-7 所示的面向接口的 UML 图软件框架

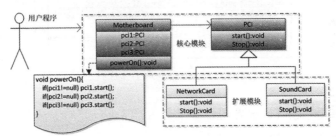

图 6-7 面向接口编程的 UML 图软件框架(A)

使用 Eclipse 创建 Java 项目"task6_9",在该项目中创建一个名为"Task6_3"的 Java 类。Task6_3.java 文件中的代码如图 6-8 所示(其中包含了 PCI 接口、使用 PCI 接口的 MotherBoard 类,以及实现 PCI 接口的 NetworkCard 类和 SoundCard 类的代码)。

```java
interface PCI{
    public abstract void start();      //声明PCI接口
    public abstract void stop();       //声明抽象方法
}                                      //声明抽象方法
class MotherBoard{
    PCI pci1,pci2,pci3;                //把对象成员声明为接口类型
    void PowerOn(){
        if(pci1!=null) pci1.start();
        if(pci2!=null) pci2.start();
        if(pci3!=null) pci3.start();
    }
    void ShutDown(){
        if(pci1!=null) pci1.stop();
        if(pci2!=null) pci2.stop();
        if(pci3!=null) pci3.stop();
    }
}
class NetworkCard implements PCI{
    public void start(){
        System.out.println("网卡开始工作!");
    }
    public void stop(){
        System.out.println("网卡停止工作!");
    }
}
class SoundCard implements PCI{
    public void start(){
        System.out.println("声卡开始工作!");
    }
    public void stop(){
        System.out.println("声卡停止工作!");
    }
}
public class Task6_9 {
    public static void main(String[] args) {
        MotherBoard mBoard = new MotherBoard();
        mBoard.pci1 = new NetworkCard();
        mBoard.pci2 = new SoundCard();
        mBoard.PowerOn();
        mBoard.ShutDown();
    }
}
```

图 6-8 Task6_9.java 文件代码

（八）面向接口编程 B

实验内容：通过实验理解面向接口编程的基本思想（把方法的参数定义为接口变量）编程实现图 6-9 所示的面向接口的 UML 图软件框架

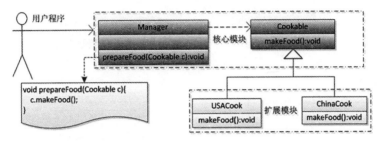

图 6-9　面向接口编程的 UML 图软件框架（B）

使用 Eclipse 创建 Java 项目"task6_10"，在该项目中创建一个名为"Task6_10"的 Java 类。Task6_10.java 文件中的代码如图 6-10 所示（其中包含了 Cookable 接口、使用 Cookable 接口的 Manager 类，以及实现 Cookable 接口的 ChinaCook 类和 USACook 类的代码）。

```java
interface Cookable{                              //定义接口
    public abstract void makeFood();             //抽象方法
}
class Manager{
    public void prepareFood(Cookable c){         //把接口作为参数
        c.makeFood();
        System.out.println("工人兄弟们，饭已OK，可以吃了！");
    }
}
class USACook implements Cookable{
    public void makeFood(){                      //用普通方法实现n阶乘
        System.out.println("I have made much USA food for you!");
    }
}
class ChinaCook implements Cookable{
    public void makeFood(){                      //用普通方法实现n阶乘
        System.out.println("我做了很多中国美食！");
    }
}
public class Task6_10 {
    public static void main(String[] args) {
        Manager a = new Manager();
        a.prepareFood(new USACook());
        Manager b = new Manager();
        b.prepareFood(new ChinaCook());
    }
}
```

图 6-10　Task6_10.java 文件代码

（九）实现多个接口

实验内容：通过实验掌握在一个类中实现多个接口的方法编程实现图 6-11 所示的 UML 图接口和实现这些接口的类

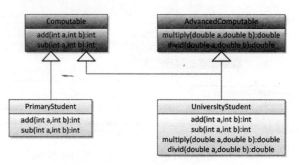

图 6-11　实现多个接口的 UML 图

使用 Eclipse 创建 Java 项目"task6_11"，在该项目中创建一个名为"Task 6_11"的

```java
interface Computable{
    public abstract int add(int a,int b);    //抽象方法，计算a+b;
    public abstract int sub(int a,int b);    //抽象方法，计算a-b;
}
interface AdvancedComputable{
    public abstract double multiply(double a,double b);  //抽象方法，计算a*b;
    public abstract double divid(double a,double b);     //抽象方法，计算a/b;
}
class PrimaryStudent implements Computable{
    public int add(int a,int b){
        return a + b;
    }
    public int sub(int a,int b){
        return a - b;
    }
}
class University implements Computable,AdvancedComputable{
    public int add(int a,int b){
        return a + b;
    }
    public int sub(int a,int b){
        return a - b;
    }
    public double multiply(double a,double b){
        return a * b;
    }
    public double divid(double a,double b){
        return a / b;
    }
}
public class Task6_11 {
    public static void main(String[] args) {
        University u = new University();
        Computable c = u;
        AdvancedComputable ac = u;
        //问题1：语句"System.err.println(u.add(10,20));"是否合法？
        //问题2：语句"System.err.println(u.multiply(3,4));"是否合法？
        //问题3：语句"System.err.println(c.add(3,4));"是否合法？
        //问题4：语句"System.err.println(c.multiply(3,4));"是否合法？
        //问题5：语句"System.err.println(ac.sub(5,2));"是否合法？
        //问题6：语句"System.err.println(ac.divid(6,3));"是否合法？
    }
}
```

图 6-12　Task 6_11.java 文件代码

Java 类。Task6_11.java 文件中的代码如图 6-12 所示(其中包含了 Computable 接口和 AdvancedComputable 接口,以及实现 Computable 接口的 PrimaryStudent 类和同时实现了 Computable 接口与 AdvancedComputable 接口的 University 类的代码)。

回答图中代码里的 6 个问题,答案直接写在图 6-12 实验作业界面空白处。

(十) 面向接口编程实践

实验内容:进一步理解面向接口编程的基本思想,编程实现图 6-13 所示的面向接口的 UML 图软件框架

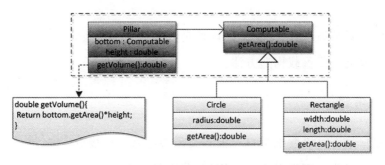

图 6-13 面向接口编程的 UML 图软件框架

使用 Eclipse 创建 Java 项目"task6_12",在该项目中创建一个名为"Task6_12"的 Java 类。Task6_12.java 文件中的代码包含了 Computable 接口、使用 Computable 接口的 Pillar 类,以及实现 Computable 接口的 Circle 类和 Rectangle 类(Task6_12 主类的 main 方法中的代码如图 6-14 所示,其他代码略,可参考 task6_9)。

```
public static void main(String[] args) {
    Circle c1 = new Circle();
    c1.r = 10;
    Pillar p1 = new Pillar();
    p1.height = 10;
    p1.bottom = c1;
    System.out.println(p1.getVolume());
    Rectangle r1 = new Rectangle();
    r1.width = 4;
    r1.length = 5;
    p1.bottom = r1;
    System.out.println(p1.getVolume());
}
```

图 6-14 Task6_12.java 的 main 方法代码

（十一）继承与接口

1. 设计一个父类 Father

(1) 成员变量：name,nationality,age,gender。

(2) 成员方法：introduce(),walk(),speak()。

2. 设计一个子类 Child

(1) 成员变量：【name,nationality,age,gender,hobby】。

(2) 成员方法：【introduce(),walk(),speak(),dancing(),singing()】,其中 introduce()要求方法重写实现。

3. 要求

(1) 设计好类后,创建对象验证设计是否满足要求。

(2) 所有类需要重载构造方法。

四、思考题

(1) 接口是否可以继承接口？抽象类是否可以实现接口？

(2) Java 异常处理涉及 5 个关键字,分别是什么？

实验七　常用系统类的使用

一、实验目的

● 了解Java常用的系统类,包括Java Applet、字符串类、输入/输出流类、数学函数类、日期类、随机数类以及向量类等的基本使用方法。

● 理解Java系统类的构成。

二、实验要求

● 进一步了解Applet类。

● 掌握不同类型的输入/输出流类、标准数据流、文件流、数据输入/输出流、对象流等。

● 掌握数学函数类的使用方法。

● 掌握日期类的使用方法。

● 掌握向量类的使用方法。

三、实验内容

(一) 了解Applet的生命周期

1. 编写KY7_1.java程序文件

源代码如下。

```java
import java.applet.Applet;
import java.awt.Graphics;
public class KY7_1 extends Applet {
    StringBuffer buffer=new StringBuffer();
    public void init() {
        addWords("执行了初始化方法 init()...");
    }
    public void start() {
        addWords("执行了开始方法 start()...");
    }
    public void stop() {
        addWords("执行了停止方法 stop()...");
    }
    public void destroy() {
        addWords("执行了清除方法 destroy()...");
    }
    void addWords(String s) {
        System.out.println(s);
        buffer.append(s);
        repaint();
    }
    public void paint(Graphics g) {
        g.drawString(buffer.toString(),5,15);
    }
}
```

2. 编译 KY7_1.java 文件

编译该文件。

3. 编写显示 KY7_1.class 的页面文件 KY7_1.html

代码如下。

<html>

<applet   codebase=e：/java/程序 code=KY7_1.class width=300 height=120>

</applet>

</html>

4. 在命令提示符窗口调用小程序查看器浏览 KY7_1.html 页面观察 Applet 应用程序

(二) 使用数学函数类

Math 是一个最终类,含有基本数学运算函数,如指数运算、对数运算、求平方根、三角函数、随机数等,可以直接在程序中加"Math."前缀调用。

1. 创建使用 Math 类的应用程序 KY7_2.java

源程序如下。

```java
import java.util.*;
class KY7_2 {
    public static void main(String args[]) {
        Random r1=new Random(1234567890L);
        Random r2=new Random(1234567890L);
        boolean b=r1.nextBoolean();  //随机数不为 0 时取真值
        int i1=r1.nextInt(100);  //产生大于等于 0 且小于 100 的随机数
        int i2=r2.nextInt(100);  //同上
        double i3=r1.nextDouble();  //产生大于等于 0.0 且小于 1.0 的随机数
        double i4=r2.nextDouble();  //同上
        double d1=Math.sin(Math.toRadians(30.0));
        double d2=Math.log(Math.E);
        double d3=Math.pow(2.0, 3.0);
        int r=Math.round(33.6F);
        System.out.println("b 的随机数不为 0 时 "+b);
        System.out.println("i1 的随机数为 "+i1);
        System.out.println("i2 的随机数为 "+i2);
        System.out.println("d1 的随机数为 "+i3);
        System.out.println("d2 的随机数为 "+i4);
        System.out.println("30 弧度的正弦值:Math.sin(Math.toRadians(30.0))= "+d1);
        System.out.println("E 的对数值:Math.log(Math.E)= "+d2);
        System.out.println("2 的 3 次方:Math.pow(2.0, 3.0)= "+d3);
```

```
        System.out.println("33.6F 四舍五入：Math.round(33.6F)= "+r);
    }
}
```

2. 编译文件

编译 KY7_3.java 文件。

3. 运行文件

在命令提示符窗口运行 KY7_3.class 文件，并输入 1234。

（三）使用日期类

Java 提供了 3 个日期类：Date,Calendar 和 DateFormat。其中，Date 类主要用于创建日期对象并获取日期，Calendar 类可获取和设置日期，DateFormat 类用来设置日期的格式。

Java 程序设计规定的基准日期为 1970.1.1 00：00：00 格林及治（GMT）标准时间，当前日期是由基准日期开始所经历的毫秒数转换出来的。

1. 使用日期类的 Applate 应用程序

（1）程序功能：说明 3 个日期类 Date、Calendar 和 DateFormat 的使用方式及显示的样式。

（2）编写 KY7_4.java 程序文件，源代码如下。

```
import java.text.*;
import java.util.*;
import java.awt.*;
import java.applet.*;
public class KY7_4 extends Applet {
  public void paint(Graphics g) {
    Date today;
    Calendar now;
    DateFormat f1,f2;
    String s1,s2;
    today=new Date();  //获取系统当前日期
    g.drawString("字符串格式："+today.toString(),20,20);
```

```
    f1=DateFormat.getInstance();//以默认格式生成格式化器
    s1=f1.format(today);//将日期转换为字符串
    g.drawString("系统格式："+s1,20,40);
    //生成长格式的中国日期格式化器
    f1=DateFormat.getDateInstance(DateFormat.LONG, Locale.CHINA);
    //生成长格式的中国时间格式化器
    f2=DateFormat.getTimeInstance(DateFormat.LONG, Locale.CHINA);
    s1=f1.format(today);//将日期转换为日期字符串
    s2=f2.format(today);//将日期转换为时间字符串
    g.drawString("中国格式："+s1+" "+s2,20,60);
    now=Calendar.getInstance();//获取系统时间
    s1=now.get(now.HOUR)+"时"+now.get(now.MINUTE)+"  分"
    +now.get(now.SECOND)+"秒";
    g.drawString("调整前时间："+s1,20,80);
    now.set(2004,8,15,9,9,9);
    today=now.getTime();
    g.drawString("调整后时间："+today.toString(),20,100);
    }
}
```

（3）编译程序文件。

（4）在浏览器中打开包含应用程序的页面文件。

2．在独立运行的应用程序中使用日期函数

（1）程序功能：补充说明 3 个日期类 Date、Calendar 和 DateFormat 的使用方式及显示样式。

（2）编写 KY7_5.java 程序文件,源代码如下。

```
import java.util.*;
import java.text.*;
public class KY7_5
{
    public static void main (String args[])
    {
```

```
Date today = new Date();  //当前日期和时间
SimpleDateFormat sdf;
sdf = new SimpleDateFormat("yyyy 年 MM 月 dd 日 hh 时 mm 分 ss 秒 a EEEEE");
System.out.println("当前日期和时间:"+sdf.format(today));
long hms=System.currentTimeMillis();  //当前时间的毫秒数
System.out.println("当前时间的毫秒数="+hms);
Date tomorrow = new Date(hms+24*60*60*1000);
System.out.println("明天是"+sdf.format(tomorrow));
Calendar now = Calendar.getInstance();
int year =now.get(Calendar.YEAR);  //年份
int month=now.get(Calendar.MONTH)+1;  //月份
int day = now.get(Calendar.DATE);  //日期
System.out.print("今天是"+year+"年"+month+"月"+day+"日");
int week = now.get(Calendar.DAY_OF_WEEK);  //星期
switch(week)
{
    case 1:System.out.println("星期日");break;
    case 2:System.out.println("星期一");break;
    case 3:System.out.println("星期二");break;
    case 4:System.out.println("星期三");break;
    case 5:System.out.println("星期四");break;
    case 6:System.out.println("星期五");break;
    case 7:System.out.println("星期六");break;
    }
  }
}
```

(3) 编译并运行程序。

四、思考题

(1) 字符串的连接符是什么？

(2) 集合类和数组类在使用上有什么区别？

实验八　图形用户界面与多媒体

一、实验目的

● 了解图形用户界面基本组件窗口、按钮、文本框、选择框、滚动条等的使用方法。

● 了解如何使用布局管理器对组件进行管理,以及如何使用 Java 的事件处理机制。

● 熟悉图形、图像的使用方法,理解计算机动画的原理和 Java 的多线程处理机制,能够编写 Applet 中使用的动画。

二、实验要求

● 掌握在 Applet 容器中添加组件的方法,掌握使用布局管理器对组件进行管理的方法。

● 理解 Java 的事件处理机制,掌握为不同组件编写事件处理程序的方法。

● 掌握编写独立运行的窗口界面的方法。

● 了解 Java Swing 组件的使用方法。

● 了解对话框的使用方法。

● 掌握使用图形类 Graphics 画出不同图形的方法。

● 掌握在容器中输入图像、播放音乐的方法。

- 理解计算机动画原理,掌握图形双缓冲技术,能够设计计算机动画。
- 理解多线程机制,掌握线程的使用方法。

三、实验内容

(一)创建图形用户界面

图形用户界面(Graphic User Interface,简称 GUI)是为方便用户使用设计的窗口界面,在图形用户界面中用户可以看到什么就操作什么,取代了在字符方式下知道是什么后才能操作的方式。组件(Component)是构成 GUI 的基本要素,通过对不同事件的响应来完成和用户的交互或组件之间的交互。组件一般作为一个对象放置在容器(Container)内,容器是能容纳和排列组件的对象,如 Applet、Panel(面板)、Frame(窗口)等。通过容器的 add 方法把组件加入到容器中。

1. 在 Applet 中添加标签、按钮并使用网格布局

(1)程序功能:在 Applet 容器中添加组件标签、按钮,并使用网格布局管理器排列组件在容器中的位置。

(2)编写 KY8_1.java 程序文件,源代码如下。

```
import java.awt.*;
import java.applet.Applet;
public class KY8_1 extends Applet{
    Label l1;
    Button b1, b2, b3, b4, b5, b6;
    public void init() {
        setLayout(new GridLayout(3,3));  //设置网格布局(3 行 3 列共 9 个网格)
        l1=new Label("标签 1 ");
        b1 = new Button("按钮 1 ");
        b2 = new Button("按钮 2 ");
        b3 = new Button("按钮 3 ");
        b4 = new Button("按钮 4 ");
        add(l1);
        add(b1);
        add(b2);
```

```
        add(b3);
        add(new Label());
        add(b4);
        add(new Button("按钮 5 "));
        add( new Button("按钮 6 "));
        add(new Label("标签 2 "));
    }
}
```

(3) 编译程序 KY8_1.java。

(4) 编写显示 Applet 的页面文件 KY8_1.html。

2. 在面板中添加组件

(1) 程序功能：在 Applet 中添加面板容器，并分别在 Applet、面板容器中添加组件并使用不同的布局管理方式。

(2) 编写 KY8_2.java 程序文件，源代码如下。

```
import java.awt.*;
import java.awt.Color;
import java.applet.Applet;
public class KY8_2 extends Applet {
    public void init() {
        //设置最底层的 Applet 容器为顺序布局
        setFont(new Font("Arial",Font.PLAIN,20));
        Label l＝new Label("这是最底层的 Applet 容器中的标签",Label.CENTER);
        add(l);
        Panel panel1＝new Panel();
        add( panel1);
        panel1.setBackground(Color.blue);
        panel1.setForeground(Color.red);
        panel1.setLayout(new BorderLayout());//设置边界布局
        panel1.add("North", new Button("北"));
        panel1.add("South", new Button("南"));
        panel1.add("East", new Button("东"));
```

```
        panel1.add("West", new Button("西"));
        panel1.add("Center", new Label("这是在 Panel1 面板中部添加的标签"));
        Panel panel2=new Panel();
        add( panel2);
        panel2.setLayout(new GridLayout(3,1)); //设置网格布局
        Choice c=new Choice ();//创建下拉式列表
        c.addItem("北京");
        c.addItem("上海");
        c.addItem("天津");
        Label l1=new Label("这是在 Panel2 面板中的标签");
        Button b1=new Button("Panel2 中的按钮");
        panel2.setBackground(Color.green);
        panel2.add(l1);
        100
        panel2.add(b1);
        panel2.add(c);
    }
}
```

(3) 编译程序 KY8_2.java。

(4) 编写显示 Applet 的页面文件 KY8_2.html。

(二) 了解事件处理机制

在图形用户界面中,程序和用户的交互是通过组件响应各种事件来实现的。例如,用户单击了一个按钮,意味着发生了按钮的单击事件;选中下拉框中的一个选项,意味着发生了一个选项事件。在 Java 中能产生事件的组件叫作事件源,如按钮。如果希望对单击按钮事件进行处理,可给事件源(按钮)注册一个事件监听器(如包含按钮的容器),如同签订了一个委托合同,当事件源发生事件时,事件监听器就代替事件源对发生的事件进行处理,这就是所谓的委托事件处理机制。

1. 单击按钮的事件处理程序

(1) 程序功能:使用手工布局设置组件标签、按钮的位置,为按钮编写单击事

件处理方法。当用户用鼠标单击按钮时,会听到一声响声。

(2) 编写 KY8_3.java 程序文件,源代码如下。

```
import java.awt.*;
import java.awt.event.*;
import java.applet.Applet;
//实现动作事件监听接口
public class KY8_3 extends Applet implements ActionListener {
  public void init() {
    setLayout(null);//关闭默认的顺序管理布局
    Label l=new Label("按一下按钮可听到响声! ", Label.CENTER);
    add(l);
    l.setBounds(40,10,150,30);
    Button b=new Button("按钮");
    add(b);
    b.setBounds(60,50,60,40);
    b.addActionListener (this); //注册事件源的动作监听者
  }
  public void actionPerformed(ActionEvent e) {//实现单击事件接口的方法
    Toolkit.getDefaultToolkit ().beep(); //单击事件发生时做出的反应
  }
}
```

(3) 编译程序 KY7_2.java。

(4) 编写显示 Applet 的页面文件 KY8_3.html。

2. 选择复选框和单选框按钮的事件处理程序

(1) 程序功能:在 Applte 上创建复选框、单选框、文本区域、单行文本框等组件,并实现根据用户输入的十进制数,选择不同选项可转换为二、八、十六进制数。

(2) 编写 KY8_4.java 程序文件,源代码如下。

```
import java.applet.Applet;
import java.awt.*;
import java.awt.event.*;
public class KY8_4 extends Applet implements ItemListener {
```

```java
TextArea area=new TextArea(6,30);//创建文本区
String Item[]={"二进制","八进制","十六进制","十进制"};
Checkbox cb[]=new Checkbox[5];
Checkbox radio[]=new Checkbox[5];
Label l=new Label("输入十进制数");
TextField TF=new TextField(6);//创建单行文本框
public void init() {
    add(l);add(TF);
    add(area);
    add(new Label("请选择进制："));
    for(int i=0; i<4; i++) {
        cb[i]=new Checkbox(Item[i]);
        add(cb[i]);
        cb[i].addItemListener(this);
    }
    CheckboxGroup cbGroup=new CheckboxGroup();//创建单选框
    add(new Label("请选择进制："));
    for(int i=0; i<4; i++) {
        radio[i]=new Checkbox(Item[i],cbGroup,false);
        add(radio[i]);
        radio[i].addItemListener(this);
    }
}
public void itemStateChanged(ItemEvent e) {
    int x=Integer.parseInt(TF.getText());
    if (e.getItem ()=="二进制")
        area.append ("你选择的是"+e.getItem ()+ Integer.toBinaryString(x)+"\n");
    if (e.getItem ()=="八进制")
        area.append ("你选择的是"+e.getItem ()+ Integer.toOctalString(x)+"\n");
    if (e.getItem ()=="十六进制")
        area.append ("你选择的是"+e.getItem ()+Integer.toHexString(x)+"\n");
    if (e.getItem ()=="十进制")
```

```
area.append("你选择的是"+e.getItem()+x+"\n");
   }
}
```

(3) 编译程序 KY8_4.java。

(4) 编写显示 Applet 的页面文件 KY8_4.html。

(三) 建立独立运行的窗口界面并使用匿名类

最常使用的包含组件的容器是窗口,在 Java 中窗口由 Frame 类生成。

1. 创建一个窗口界面

(1) 程序功能:创建一个具有关闭功能的空白窗口。

(2) 编写 KY8_5_W.java 程序文件,源代码如下。

```
import java.awt.*;
import java.awt.event.*;
public class KY8_5_W {
    public static void main(String[] args) {
        new KY8_5_W();
    }
    KY8_5_W(){
        Frame f=new Frame("初始窗口");//创建窗口对象
        f.setSize(350,200);//设置窗口大小
        f.setVisible(true);//设置窗口是可视的
        f.addWindowListener(new WindowAdapter()){//为窗口添加窗口事件适配器
            public void windowClosing(WindowEvent e) {//关闭窗口事件的方法
                System.exit(0);
            }
        }
    }
}
```

(3) 编译并运行程序。

2. 在窗口中添加组件

(1) 程序功能:在窗口中添加组件。

(2) 编写 KY8_6.java 程序文件,源代码如下。

```java
import java.awt.*;
import java.awt.event.*;
public class KY8_6 extends Frame implements ActionListener {
    Button btn1, btn2;
    TextField f,tf1,tf2;
    TextArea Area;
    KY8_6() {
        super("添加组件的窗口");
        addWindowListener(new WindowAdapter() {
            public void windowClosing(WindowEvent e) {
                System.exit(0);
            }
        }
        setSize(350,250);//设置窗口大小
        setLocation(200,200);//设置窗口显示位置
        setFont(new Font("Arial",Font.PLAIN,12)); //设置字体
        setLayout(new FlowLayout());
        Area=new TextArea (6,40);
        tf1=new TextField(10); tf2=new TextField(10);
        btn1=new Button("显示"); btn2=new Button("退出");
        f=new TextField(20);
        add(Area); add(new Label("用户名"));
        add(tf1); add(new Label("电话"));
        add(tf2); add(f); add(btn1); add(btn2);
        tf1.addActionListener(this); tf2.addActionListener(this);
        btn1.addActionListener(this); btn2.addActionListener(this);
        show();
    }
    public static void main(String args[]) {
        new KY8_6();
    }
    public void actionPerformed(ActionEvent e) {
```

```
        if (e.getSource()==btn1)
        f.setText("你按下了""+e.getActionCommand()+""按钮");
        if (e.getSource()==tf1)
        Area.append("用户名："+tf1.getText()+"\n     ");
        if (e.getSource()==tf2)
        Area.append("电话："+tf2.getText()+"\n");
        if (e.getSource()==btn2) {
            for (int i=0; i<100000000; i++);
            dispose();//只关闭当前窗口,注销该对象
        }
    }
}
```

(3) 编译并运行程序。

3. 为窗口添加菜单

(1) 程序功能：在窗口中添加菜单栏,在菜单栏添加菜单项,并添加下拉菜单和二级菜单,通过选择菜单项可以执行不同操作,如"打开"KY8_7 类生成的窗口。

(2) 编写 KY8_7.java 程序文件,源代码如下。

```
import java.awt.*;
import java.awt.event.*;
public class KY8_7 extends Frame implements ActionListener {
    Panel p=new Panel();
    Button b=new Button("退出");
    MenuBar mb=new MenuBar();  //以下生成菜单组件对象
    Menu m1=new Menu("文件");
    MenuItem open=new MenuItem("打开");
    MenuItem close=new MenuItem("关闭");
    MenuItem exit=new MenuItem("退出");
    Menu m12=new Menu("编辑");
    MenuItem copy=new MenuItem("复制");
    MenuItem cut=new MenuItem("剪切");
    MenuItem paste=new MenuItem("粘贴");
```

```
Menu m2=new Menu("帮助");
MenuItem content=new MenuItem("目录");
MenuItem index=new MenuItem("索引");
MenuItem about=new MenuItem("关于");
KY8_7() {
    super("添加菜单的窗口");
    setSize(350,200);
    add("South",p);
    p.add(b);
    b.addActionListener(this);
    m1.add(open); //将菜单项加入到菜单m1中
    m1.add(close);
    m1.addSeparator(); //在菜单中添加分隔条
    m1.add(exit);
    open.addActionListener(this); //给菜单项open注册事件监听器
    exit.addActionListener(this);
    mb.add(m1); //将菜单m1加入到菜单栏mb中
    m12.add(copy); m12.add(cut); m12.add(paste);
    m1.add(m12); //将m12作为二级菜单添加到m1菜单项中
    m2.add(content); m2.add(index); m2.addSeparator(); m2.add(about);
    mb.add(m2);
    setMenuBar(mb); //设置菜单栏为mb
    show(); //显示组件
}
public static void main(String args[]) {
    new KY8_7();
}
public void actionPerformed(ActionEvent e) {
    if (e.getActionCommand()=="退出")
        System.exit(0);
    if (e.getActionCommand()=="打开")
```

```
        new KY8_6();
    }
}
```

(3) 编译并运行程序。

(四) 使用 Swing 组件

在 Java 中,能够实现图形用户界面的类库有两个:java.awt 和 javax.swing。前者称为抽象窗口工具库(Abstract Windows Toolkit,AWT);后者是 Java 基础类库(Java Foundation Classes,JFC)的一个组成部分,它提供了一套功能更强、数量更多、更美观的图形用户界面组件。Swing 组件名称和 AWT 组件名称基本相同,但以 J 开头,例如 AWT 按钮类的名称是 Button,在 Swing 中的名称则是 JButton。

1. 在 JApplet 中添加 Swing 组件

(1) 程序功能:在 JApplet 中添加 3 个带有图片的按钮和 1 个带有图片的标签。

(2) 准备图片文件:在当前目录下建立 1 个名为 image 的文件夹,存放 4 个图片文件,例如 PreviousArrow.gif。

(3) 编写 KY8_8.java 程序文件,源代码如下。

```
import javax.swing.*;
import java.awt.*;
import java.awt.Color;
public class KY8_8 extends JApplet {
    Container pane;
    JPanel panel1,panel2;
    JButton button1,button2,button3;
    JLabel label;
    public void init() {
        pane=getContentPane();
        panel1=new JPanel(new FlowLayout());
        panel2=new JPanel(new FlowLayout());
        ImageIcon icon = new ImageIcon(" image/PreviousArrow.gif "," ");
```

```
        button1=new JButton(icon);
        button2=new JButton(new ImageIcon("image/go.GIF"));
        button3=new JButton(new ImageIcon("image/NextArrow.gif"));
        label=new JLabel("图像标签",ImageIcon("image/Candl02.gif"),SwingConstants.CENTER);
        pane.setBackground(new Color(255,255,200));
        panel1.setBackground(new Color(255,255,104));
        panel2.setBackground(new Color(255,255,214));
        button1.setToolTipText("向上翻页按钮");
        button2.setToolTipText("跳转按钮");
        button3.setToolTipText("向下翻页按钮");
        pane.add("North",panel1);
        pane.add(panel2,BorderLayout.SOUTH);
        panel1.add(button1);
        panel1.add(button2);
        panel1.add(button3);
        panel2.add(label);
    }
}
```

(4) 编译 KY8_8.java。

(5) 编写显示 KY8_8.class 的页面文件。

(6) 使用 appletviewer 查看程序结果。

2. 在 JFrame 窗口中添加组件

(1) 程序功能：创建 JFrame 窗口，并在其中添加工具栏。

(2) 准备图片文件：在当前目录下建立一个名为 image 的文件夹，存放 3 个图片文件，例如 PreviousArrow.gif。

(3) 编写 KY8_9.java 程序文件，源代码如下。

```
import javax.swing.*;
import java.awt.*;
import java.awt.event.*;
public class KY8_9 extends JFrame implements ActionListener{
```

图形用户界面与多媒体 实验八

```java
JButton button1,button2,button3;
JToolBar toolBar;
JTextArea textArea;
JScrollPane scrollPane;
JPanel panel;
public static void main(String[] args) {
  new KY8_9();
}
public KY8_9() {
  super("带有工具栏按钮的窗口");
  addWindowListener(new WindowAdapter() {
    public void windowClosing(WindowEvent e) {
      System.exit(0);
    }
  }
  button1=new JButton(new ImageIcon(" image/PreviousArrow.gif "));
  button2=new JButton(new ImageIcon(" image/go.GIF "));
  button3=new JButton(new ImageIcon(" image/NextArrow.gif "));
  button1.addActionListener(this);
  button2.addActionListener(this);
  button3.addActionListener(this);
  toolBar=new JToolBar();
  toolBar.add(button1);
  toolBar.add(button2);
  toolBar.add(button3);
  textArea=new JTextArea(6,30);
  scrollPane=new JScrollPane(textArea);
  panel=new JPanel();
  setContentPane(panel);
  panel.setLayout(new BorderLayout());
  panel.setPreferredSize(new Dimension(300,150));
  panel.add(toolBar,BorderLayout.NORTH);
```

```java
      panel.add(scrollPane,BorderLayout.CENTER);
      pack();
      show();
    }
    public void actionPerformed(ActionEvent e) {
      String s="";
      if (e.getSource()==button1)
        s="左按钮被单击\n";
      else if (e.getSource()==button2)
        s="中按钮被单击\n";
      else if (e.getSource()==button3)
        s="右按钮被单击\n";
      textArea.append(s);
    }
}
```

(4) 编译 KY8_9.java。

(5) 运行 KY8_9.class。

(五) 使用自定义对话框与内部类

对话框是 GUI 中很常见的窗口对象,有着广泛的应用。对话框和普通窗口最大的不同就是对话框是依附在某个窗口上的,一旦它所依附的窗口关闭了,对话框也要随之关闭。Java 提供了 Dialog 类用于制作自定义对话框,当需要改变一些数据或需要一个提示窗口时可使用自定义对话框。

1. 程序功能

创建一个带有文本区及"对话框"按钮的父窗口,单击"对话框"按钮可打开一个自定义对话框,从中可以定义行和列的数值,关闭对话框后其设置的数值会显示在父窗口的文本区中。

2. 编写 KY8_10.java 程序文件

源代码如下。

```java
import javax.swing.*;
import java.awt.*;
```

```java
import java.awt.event.*;
public class KY8_10 extends JFrame implements ActionListener {
    int row=10, col=40;
    JPanel p1=new JPanel(), p2=new JPanel();
    JTextArea ta=new JTextArea("文本区行数:"+row+"列数:"+col, row, col);
    JScrollPane scrollPane=new JScrollPane(ta);
    Button exit=new Button("关闭");
    Button dialog=new Button("对话框");
    JPanel panel=new JPanel();
    KY8_10() {
        setContentPane(panel);
        setTitle("带有对话框的父窗口");
        panel.setPreferredSize(new Dimension(500,200));
        panel.setLayout(new BorderLayout());
        panel.add("Center", p1); panel.add("South", p2);
        p1.add(scrollPane);
        p2.add(exit); p2.add(dialog);
        exit.addActionListener(this);
        dialog.addActionListener(this);
        pack();
        show();
        //setVisible(true);
    }
    public static void main(String args[]) {
        new KY8_10();
    }
    public void actionPerformed(ActionEvent e) {
        if (e.getSource()==exit)
            System.exit(0);
        else {
            MyDialog dlg=new MyDialog(this, true);
            dlg.show();
```

```
        }
    }
class MyDialog extends Dialog implements ActionListener {
    Label label1=new Label("请输入行数");
    Label label2=new Label("请输入列数");
    TextField rows=new TextField(50);
    TextField columns=new TextField(50);
    Button OK=new Button("确定");
    Button Cancel=new Button("取消");
    MyDialog(KY8_10 parent, boolean modal) {
        super(parent,modal);
        setTitle("自定义对话框");
        setSize(260,140);
        setResizable(false);
        setLayout(null);
        add(label1);
        add(label2);
        label1.setBounds(50,30,65,20);
        label2.setBounds(50,60,65,20);
        add(rows);
        add(columns);
        rows.setText(Integer.toString(ta.getRows()));
        columns.setText(Integer.toString(ta.getColumns()));
        rows.setBounds(120,30,90,20);
        columns.setBounds(120,60,90,20);
        add(OK);
        add(Cancel);
        OK.setBounds(60,100,60,25);
        Cancel.setBounds(140,100,60,25);
        OK.addActionListener(this);
        Cancel.addActionListener(this);
    }
```

```java
    public void actionPerformed(ActionEvent e) {
      if(e.getSource()==OK) {
        int row=Integer.parseInt(rows.getText());
        int col=Integer.parseInt(columns.getText());
        ta.setRows(row);
        ta.setColumns(col);
        ta.setText("文本区行数："+row+" 列数："+col);
        show();
      }
      dispose();
    }
  }
}
```

（3）编译并运行程序。

（六）使用图形类

在 Java 中基本图形包括点、线、圆、矩形等，是构成复杂图形的基础。绘制基本图形要使用 AWT 包中的 Graphics 类，它提供了各种基本图形的绘制方法，可以直接引用这些方法画点、线、圆、矩形等。

1. 创建在 Applet 上画出不同的图形的程序

（1）程序功能：在 Applet 上使用不同的颜色画出直线、圆、方块、圆弧等图形。

（2）编写 KY8_11.java 程序文件，源代码如下。

```java
import java.applet.Applet;
import java.awt.Graphics;
import java.awt.Color;
public class KY8_11 extends Applet {
  public void paint(Graphics g) {
    g.drawLine(10,10,50,10);//画线（确定两点）
    g.setColor(Color.red);//设置红颜色
    g.drawOval(35,35,100,60);//画椭圆（圆心、宽和高）
    g.fillOval(200,15,60,100);//画具有填充色的圆
```

```
            g.setColor(Color.blue);//设置蓝颜色
            g.drawRect(20,130,80,80);//画矩形
            g.fillRect(120,130,80,80);//画具有填充色的矩形
            g.drawRoundRect(220,130,80,80,20,20);//画圆角矩形
            g.fillRoundRect(320,130,80,80,20,20);//画具有填充色的圆角矩形
            g.setColor(new Color(255,255,0)); //设置黄颜色
         g.drawArc (250,20,100,100,0,90);
          g.fillArc (380,20,100,100,90,90);
          g.fillArc (300,25,100,100,180,90);
         g.drawArc (330,25,100,100,0,-90);
      }
}
```

(3) 编译程序 KY8_11.java。

(4) 编写显示 KY8_11.class 的页面文件,在浏览器中显示结果。

2. 创建使用画布对象的 Applet 应用程序

(1) 程序功能:创建一个带有多边形、圆的自定义画布类,在 Applet 上显示自定义画布的对象。

(2) 编写 KY8_12.java 程序文件,源代码如下。

```java
import java.applet.Applet;
import java.awt.*;
import java.awt.Color;
public class KY8_12 extends Applet {
  public void init() {
    Color col=new Color(20,55,75);
    setBackground(col);//设置 Applet 的背景色
    setForeground(Color.yellow);//设置 Applet 的前景色
    MyCanvas1 c=new MyCanvas1();//创建画布对象
    c.setBackground(Color.white);//设置画布的背景色
    c.setSize(300,200);//设置画布的大小
    add(c);
  }
```

```
    }
class MyCanvas1 extends Canvas {
    public void paint(Graphics g) {
        g.setColor(Color.red);
        g.fillOval(40,20,80,80);//画圆
        g.setColor(Color.cyan);//设置青色
        int p1X[]={20,20,100,20};//多边形的 X 坐标
        int p1Y[]={20,80,20,20};//多边形的 Y 坐标
        int p1=3;//多边形的边数
        g.fillPolygon (p1X,p1Y,p1);//画填充多边形
        int p2X[]={280,120,50,90,210,280};
        int p2Y[]={20,50,100,110,70,20};
        int p2=5;
        g.drawPolygon (p2X,p2Y,p2);//画多边形
    }
}
```

(3) 编译程序 KY8_12.java。

(4) 编写显示 KY8_12.class 的页面文件,在浏览器中显示结果。

(七) 插入图像与播放音乐

1. 在 Applet 中插入图像播放音乐

(1) 程序功能:在 Applet 中插入 3 种大小的图像,并在打开文件时播放背景音乐。

(2) 编写 KY8_13.java 程序文件,源代码如下。

```
import java.awt.*;
import java.applet.*;
public class KY8_13 extends Applet {
    Image img;
    public void init(){
        //获取图像文件地址读取图像文件到内存
        img=getImage(getCodeBase(),"image/飞机.gif");
        play(getDocumentBase(),"WAV/Sound.wav");      //播放声音文件
```

```
        }
        public void paint(Graphics g){
            int w=img.getWidth(this);
            int h=img.getHeight(this);
            g.drawImage(img,20,10,this);                    //画出原图
            g.drawImage(img,20,100, w/2, h/2, this);        //画出缩小一半的图
            g.drawImage(img,160,0, w*2, h*2, this);         //画出放大一倍的图
        }
    }
```

（3）编译程序 KY8_13.java。

（4）编写显示 KY8_13.class 的页面文件。

2．随时播放声音文件的程序

（1）程序功能：在 Applet 中使用下拉框显示音乐文件的名字，在程序中使用 AudioClip 类控制播放的音乐文件，使用"播放""连续""停止"按钮控制和选择播放音乐的状态。

（2）编写 KY8_14.java 程序文件，源代码如下。

```
        import java.awt.*;
        import java.awt.event.*;
        import java.applet.Applet;
        import java.applet.AudioClip;
        public class KY8_14 extends Applet implements ItemListener, ActionListener {
            AudioClip sound;
            Choice c=new Choice();
            Button play=new Button("播放");
            Button loop=new Button("连续");
            Button stop=new Button("停止");
            public void init() {
                c.add("space.au"); c.add("flute.aif"); c.add("trip.mid");
                c.add("jungle.rmf"); c.add("Sound.wav");
                add(c); c.addItemListener(this);
                add(play); add(loop); add(stop);
```

```
        play.addActionListener(this);
        loop.addActionListener(this);
        stop.addActionListener(this);
        sound=getAudioClip(getCodeBase(),"WAV/Sound.wav");
    }
    public void itemStateChanged(ItemEvent e) {
        sound.stop();
        sound=getAudioClip(getCodeBase(),"WAV/"+c.getSelectedItem());
    }
    public void actionPerformed(ActionEvent e) {
        if (e.getSource()==play) sound.play();
        else if (e.getSource()==loop) sound.loop();
        else if (e.getSource()==stop) sound.stop();
    }
}
```

(3) 编译程序 KY8_14.java。

(4) 编写显示 KY8_14.class 的页面文件，在浏览器中显示结果。

(八) 幻灯机效果——连续显示多幅图像

1. 程序功能

如果 Applet 仅仅是显示一幅图像，没有什么特别的意义，不如直接在 HTML 文件中显示图像。本程序可以像幻灯机那样连续显示多幅图像。

2. 准备图像文件

在当前目录中的 image 文件夹中准备 6 幅花的图像文件。

3. 编写 KY8_15.java 程序文件

源代码如下。

```
import java.awt.*;
import java.awt.event.*;
import java.applet.*;
public class KY8_15 extends Applet {
    int index;
    Image imgs[]=new Image[6];
```

```
    public void init(){
      addMouseListener(new MouseAdapter() {
        public void mouseClicked(MouseEvent e) {
          index=++index%6;
          repaint();
        }
      }
      for (int i=0; i<6; i++)
        imgs[i]=getImage(getCodeBase(),"image/花"+(i+1)+".gif");
    }
    public void paint(Graphics g){
      if (imgs[index]!=null)
        g.drawImage(imgs[index],60,20,this);
    }
}
```

在这个程序中,加载了 6 幅图像,单击鼠标可逐一显示图像,并在显示完 6 幅图像后自动返回第一幅重新开始。

对程序进行分析,写出分析结果。

(九) 使用滚动条改变背景颜色

1. 程序功能

移动滚动条可以改变背景颜色。

2. 编写 KY8_16.java 程序文件

源代码如下。

```
import java.applet.Applet;
import java.awt.*;
import java.awt.event.*;
import java.awt.Scrollbar;
import java.awt.Color;
public class KY8_16 extends Applet implements AdjustmentListener {
  Scrollbar r1,r2,r3;
  int red,green,blue;
```

```java
TextField t;Label a;
public void init() {
    setLayout(null);
    r1=new Scrollbar(Scrollbar.HORIZONTAL,0,1,0,255);
    r2=new Scrollbar(Scrollbar.HORIZONTAL,0,1,0,255);
    r3=new Scrollbar(Scrollbar.HORIZONTAL,0,1,0,255);
    t=new TextField("0",5);
    t.setEditable(false);
    a=new Label("移动滚动条可改变背景颜色",Label.CENTER);
    add(a);a.setBounds(120,10,150,15);
    add(r1);r1.setBounds(20,30,100,20);
    add(r2);r2.setBounds(140,30,100,20);
    add(r3);r3.setBounds(260,30,100,20);
    add(t);t.setBounds(20,120,220,18);
    r1.addAdjustmentListener(this);
    r2.addAdjustmentListener(this);
    r3.addAdjustmentListener(this);
}
public void adjustmentValueChanged(AdjustmentEvent e) {
    red=r1.getValue();
    green=r2.getValue();
    blue=r3.getValue();
    t.setText("red 的值"+String.valueOf(r1.getValue())+
    ",green 的值"+String.valueOf(r2.getValue())+",blue 的值"+
    String.valueOf(r3.getValue()));
    Color c=new Color(red,green,blue);
    setBackground(c);
}
}
```

3．分析并写出结果

对程序进行分析，写出分析结果。

（十）Applet 与 Application 合并运行

Java Applet 和 Application 程序的区别在于运行方式不同。那么能不能将它们合并起来，让同一个程序既可以由浏览器运行，又可以单独运行呢？

1. 程序功能

在 Applet 与 Application 方式下都能运行。

2. 编写 KY8_17.java 程序文件

源代码如下。

```java
import java.applet.*;
import java.awt.*;
import java.awt.event.*;
public class KY8_17 extends Applet implements ActionListener {
    Button button;
    TextField field;
    public static void main(String[] args) {
        Frame window=new Frame("AppDemo");//创建窗口对象
        AppDemo app=new AppDemo();//创建程序对象
        window.add("Center",app);//将程序对象添加到窗口
        app.init();//调用程序的初始化方法
        window.addWindowListener(new WindowAdapter() {
            public void windowClosing(WindowEvent e) {
                System.exit(0);
            }
        });//以上用匿名类的方式为窗口添加关闭功能
        window.setSize(300,120);//设定窗口大小
        window.setVisible(true);//设定窗口可见
    }
    public void init() {
        button=new Button("显示");
        button.addActionListener(this);
        field=new TextField(23);
        add(field);
```

```
        add(button);
     }
     public void actionPerformed(ActionEvent e){
        field.setText("Applet 与 Application 的合并运行");
     }
  }
```

3. 编译程序

编译 KY8_18.java 源程序。

4. 编写浏览 Applet 的页面文件

编写浏览 Applet 的页面文件 KY8_18.html,在浏览器打开文件 KY8_18.htm。

5. 运行文件

在独立运行的 Application 方式下运行 KY8_18.class 字节文件。

(十一)创建电闪雷鸣的动画

1. 程序功能

本程序可以通过按钮控制声音和动画的开始和停止,动画显示了电闪雷鸣的场面。注意:图像文件要分别表现不同时间段的电闪场面,这样才会有动画效果。

2. 编写 KY8_18.java 程序文件

源代码如下。

```
import java.awt.*;
import java.applet.*;
import java.awt.event.*;
public class KY8_18 extends Applet implements Runnable,ActionListener{
    Image iImages[];                    //图像数组
    Thread aThread;
    int iFrame;                         //图像数组下标
    AudioClip au;                       //定义一个声音对象
    Button b1,b2;
    public void init(){
```

```
        int i,j;
        iFrame=0;
        aThread=null;
        iImages = new Image[10];
        for (i=0;i<10;i++){
            iImages[i]=getImage(getCodeBase(),"image/"+"tu"+(i+1)+".JPG");
        }
        au=getAudioClip(getDocumentBase(),"Wav/Sound.wav");
        au.play();                          //播放一次声音文件
        Panel p1 = new Panel();
        b1 = new Button("开始");
        b2 = new Button("停止");
        p1.add(b1);
        p1.add(b2);
        b1.addActionListener(this);
        b2.addActionListener(this);
        setLayout(new BorderLayout());
        add(p1,"South");
    }
    public void start(){
        if (aThread == null)
        {
            aThread = new Thread(this);
            aThread.start();                //线程启动
            b1.setEnabled(false);
        }
    }
    public void stop(){
      if (aThread != null) {
        aThread.interrupt();                //线程中断
        aThread = null;
        au.stop();                          //停止播放声音文件
```

```java
            }
        }
        public void run() {
            while (true)
            {
                iFrame++;
                iFrame %= (iImages.length);    //下一幅图像的下标
                repaint();
                try{
                    Thread.sleep(50);
                }
                catch (InterruptedException e)
                {                              //中断时抛出
                    break;                     //退出循环
                }
            }
        }
    public void update(Graphics g) {
        g.drawImage(iImages[iFrame],0,0,this);
    }
    public void actionPerformed(ActionEvent e) {
        if ((e.getSource()==b1) && (aThread == null))
        { //单击 Start 按钮时触发
            aThread = new Thread(this);
            aThread.start();                   //线程启动
            b1.setEnabled(false);
            b2.setEnabled(true);
            au.loop();                         //循环播放声音文件
        }
        if ((e.getSource()==b2) && (aThread != null))
        {                                      //单击 Stop 按钮时触发
            aThread.interrupt();               //线程中断
```

```
                aThread = null;
                b1.setEnabled(true);
                b2.setEnabled(false);
                au.stop();                    //停止播放声音文件
            }
        }
}
```

3. 编译源程序

编译并运行该程序。

4. 编写浏览 Applet 的页面文件

编写浏览 Applet 的页面文件,在浏览器运行结果。

四、思考题

（1）Swing 和 AWT 有什么区别？

（2）简述 Java 的事件处理机制。什么是事件源？什么是监听器？在 Java 的图形用户界面中,哪些组件可以充当事件源？

实验九　流与文件

一、实验目的

- 理解数据流的概念。
- 理解 Java 流的层次结构。
- 理解文件的概念。

二、实验要求

- 掌握字节流的基本使用方法。
- 掌握字符流的基本使用方法。
- 能够创建、读写、更新文件。

三、实验内容

（一）掌握 File 类常用方法的使用

使用 Eclipse 创建 Java 项目"task9_1"，在该项目中创建一个名为"Task9_1"的 Java 主类。在 Task9_1 类的 main 方法中，实现如下功能：程序运行后，提示"请输入一个包含路径的文件（夹）名："，用户输入一个字符串后按 Enter 键，程序判断该文件（夹）是否存在，若存在，则显示相关文件（夹）属性，否则提示"该文件（夹）不存在！"。然后再次提示"请输入一个包含路径的文件（夹）名："，再次判断该文件（夹）是否存在，并显示相关文件(夹)属性,直到用户输入的字符串为"exit"时,退出程序。程序流程如图 9-1 所示,参考代码如图 9-2 所示。程序运行效果参见 Task9_1.class(下载后保存到 D 盘根目录,并在 MS-DOS 窗口先执行命令"d:"再执行命令"java Task9_1")。

将图 9-2 中涂黑部分补充完整,从而实现如图 9-1 所示的程序流程的全部功能。

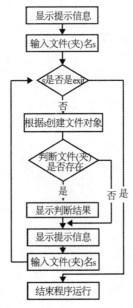

图 9-1 Task9_1 程序流程

```
Scanner reader = new Scanner(System.in);
System.out.println("请输入一个包含路径的文件(夹)名:");
String s = reader.nextLine();
while(!s.equals("exit")){
    File file =            ;     //创建文件对象
    if(file.     ){
        if(file.     ){        //判断是否是文件
            System.out.println("输入的路径是一个文件路径");
            System.out.println("该文件的绝对路径:"+file.        );
            System.out.println("该文件的父目录:"+file.        );
            System.out.println("该文件的文件名:"+file.        );
            System.out.println("该文件的大小(字节数):"+file.        );
            System.out.println("该文件是否可执行文件:"+file.        );
            System.out.println("该文件是否可写:"+file.        );
        }
        else{
            System.out.println("输入的路径是一个文件夹路径");
            System.out.println("该文件夹是否隐藏:"+file.        );
            System.out.println("该文件夹的绝对路径:"+file.        );
            System.out.println("该文件夹的父目录:"+file.        );
            System.out.println("该文件夹所在磁盘的剩余空间(字节数):"+file.        );
        }
    }
    else {
        System.out.println("该文件(夹)不存在!"+s);
    }
    System.out.println("-----------------------------------------");
    System.out.println("请输入一个包含路径的文件(夹)名:");
    s = reader.nextLine();
}
System.out.println("谢谢您的使用");
reader.close();
```

图 9-2 Task9_1.java 文件代码

（二）掌握 File 类常用方法的使用

使用 Eclipse 创建 Java 项目"task9_2"，在该项目中创建一个名为"Task9_2"的 Java 主类。Task9_2.java 文件中的代码如图 9-3 所示（其中包含了 FileFilter 类、FileOperator 类的代码）。

```java
//实现了FilenameFilter接口的类
class FileFilter implements FilenameFilter{
    private String extendName;
    public FileFilter(String extName){
        extendName="."+extName;
    }
    public boolean accept(File dir, String name) {    //接口FilenameFilter中声明的方法
        File f = new File(dir.getAbsolutePath()+"\\"+name);
        if(f.isDirectory())                           //是文件夹
            return true;
        else {
            return name.endsWith(extendName);
        }
    }
}
class FileOperator{
    //该方法用于遍历dir目录下的所有文件，返回目录dir下文件的个数
    public int traverseDirectory(File dir){
        int fileCount = 0;
        File[] files = dir.listFiles();
        for(File f: files){
            if(f.isFile()) {
                fileCount++;
                System.out.println(f.getAbsolutePath());
            }
        }
        for(File f: files){
            if(f.isDirectory()) fileCount=fileCount+traverseDirectory(f);
        }
        return fileCount;
    }
    //该方法用于遍历dir目录下的满足filter要求的所有文件，返回满足filter要求的文件个数
    public int traverseDirectory(File dir,FilenameFilter filter){  //遍历dir目录下的满足filter的所有文件
        int fileCount = 0;
        File[] files = dir.listFiles(filter);
        for(File f: files){
            if(f.isFile()) {
                fileCount++;
                System.out.println(f.getAbsolutePath());
            }
        }
        for(File f: files){
            if(f.isDirectory()) fileCount=fileCount+traverseDirectory(f,filter);
        }
        return fileCount;
    }
}
public class Task9_2 {
    public static void main(String args[]){
        FileOperator fileTraverser = new FileOperator();
        int fileCount = fileTraverser.traverseDirectory(new File("d:\\workspace"));
        System.out.println(fileCount);
        //FilenameFilter filter = new FileFilter("java");
        //fileCount = fileTraverser.traverseDirectory(new File("d:\\workspace"),filter);
        //System.out.println(fileCount);
    }
}
```

图 9-3 Task9_2.java 文件代码

运行程序，测试是否可以实现对文件夹的遍历。

删除最后 3 行代码的注释符，再次运行程序，体会文件过滤器的功能。

（三）文件字节流的使用

通过实验掌握文件字节流 FileInputStream 类和 FileOutputStream 类的使用。使

用 Eclipse 创建 Java 项目"task9_3",在该项目中创建一个名为"Task9_3"的 Java 主类。Task9_3.java 文件中的代码如图 9-4 所示(其中包含了 FileEncryption 类的代码)。

```
//文件加密类
class FileEncryption{
    private String password;
    public void setPassword(String pwd){        //设置密钥
        password = pwd;
    }
    //encodeFile方法用于加密文件,参数sFile表示需要加密的源文件,参数dFile表示需要加密后的文件
    public boolean encodeFile(String sFile,String dFile){
        File file = new File(sFile);
        if(▓▓▓▓▓▓▓▓▓){                          //如果文件不存在,则直接返回false
            return false;
        }
        byte[] pwd = password▓▓▓▓▓▓▓;           //获取密钥所对应的字节数组
        byte[] buffer = new byte[pwd.length];    //定义字节数组buffer,用于从文件字节输入流中读取数据
        try{
            FileInputStream in = ▓▓▓▓▓▓▓▓▓▓▓▓▓▓▓▓▓▓;    //创建文件输入流对象
            FileOutputStream out = new FileOutputStream(dFile,false);
            int byteCount = in.▓▓▓▓▓▓;           //从文件字节输入流中读取若干个字节到字节数组buffer中
            while(▓▓▓▓▓▓▓▓▓){                    //若读取到的字节数大于 0,则继续循环
                for(int i=0;i<byteCount;i++){    //对字节数组buffer进行异或加密
                    buffer[i]=(byte)(buffer[i]^pwd[i]);
                }
                ▓▓▓▓▓▓▓▓▓▓▓▓▓▓▓▓▓▓▓▓▓;           //把数组buffer中的前byteCount个字节写到文件字节输出流中
                byteCount = in.read(buffer);
            }
            in.close();                          //关闭流
            out.close();
            return true;
        }
        catch(IOException e){
            return false;
        }
    }
    public boolean decodeFile(String sFile,String dFile){
        return encodeFile(sFile,dFile);
    }
}
public class Task9_3 {
    public static void main(String[] args) {
        FileEncryption fileEn = new FileEncryption();    //创建加密类对象
        fileEn.setPassword("dswybs");
        fileEn.encodeFile("d:\\aa.txt","d:\\encodedaa.txt");    //加密文件
        fileEn.decodeFile("d:\\encodedaa.txt","d:\\bb.txt");    //解密文件
        System.out.println("OK");
    }
}
```

图 9-4 Task9_3.java 文件代码

将图 9-3 所示代码中被涂黑部分补充完整,从而实现文件异或加密与解密功能。

(四) 文件字符流的使用

通过实验掌握文件字符流 FileReader 类和 FileWriter 类的使用。使用 Eclipse 创建 Java 项目"task9_4",在该项目中创建一个名为"Task9_4"的 Java 主类。Task9_4.java 文件中的代码如图 9-5 所示(其中包含了 CharCounter 类的代码)。

实验九 流与文件

```java
//英文字母统计类，实现统计26个英文字母出现的次数
class CharCounter{
    private int[] CC = new int[26];      //数组CC表示26个计数器，用于统计26个英文字母出现的次数
    //方法CountChars用于统计文件fileName中26个英文字母出现的次数
    public boolean CountChars(String fileName){
        int[] temp = new int[26];
        File file = ▓▓▓▓▓▓;
        if(!file.exists()) return false;
        try{
            FileReader fileReader = ▓▓▓▓▓▓▓▓▓;    //创建文件字符输入流对象
            char[] buffer = new char[10];
            int charCount = ▓▓▓▓▓▓▓▓▓;              //从文件字符输入流读取若干个字符存入buffer中
            while(▓▓▓▓▓▓){                          //若读取的字符数大于0，则继续循环
                for(int i=0;i<charCount;i++){       //对buffer中的每个字符进行统计
                    char c = buffer[i];
                    if(c>='A' && c<='Z'){           //是大写字母的情况
                        CC[c-'A']=CC[c-'A']+1;
                    }
                    else if(c>='a' && c<='z'){      //是小写字母的情况
                        ▓▓▓▓▓▓▓▓▓;                   //相应的计数器加1（参考上面的类似代码）
                    }
                }
                charCount = fileReader.read(buffer); //从文件字符输入流读取若干个字符存入buffer中
            }
            fileReader.close();
            for(int i=0;i<CC.length;i++) CC[i]=CC[i]+temp[i];
            return true;
        }
        catch(Exception e){
            return false;
        }
    }
    //方法initialize初始化计数器数组CC
    public void initialize(){
        for(int i=0;i<CC.length;i++){
            CC[i]=0;
        }
    }
    //方法saveResult用于把统计结果保存到文件fileName中
    public boolean saveResult(String fileName){
        try{
            FileWriter writer = ▓▓▓▓▓▓▓▓▓;           //创建文件字符输出流对象
            for(int i=0;i<CC.length;i++){            //输出26个英文字母的统计情况
                writer.write("字母"+(char)(i+'A')+"的个数为："+CC[i]);
                writer.write("\r\n");
            }
            ▓▓▓▓▓▓▓▓▓;                                //关闭文件字符输出流对象
            return true;
        }
        catch(Exception e){
            return false;
        }
    }
}
public class Task9_4 {
    public static void main(String[] args) {
        CharCounter counter = new CharCounter();
        counter.initialize();
        counter.CountChars("d:\\MyCalendar.java");    //实验时，自己修改为相应的文件
        counter.saveResult("d:\\result.txt");
        System.out.println("ok");
    }
}
```

图 9-5 Task9_4.java 文件代码

将图 9-5 所示代码中被涂黑部分补充完整，从而实现统计文件中英文字母个数的功能，答案直接写在实验作业界面空白处。

（五）缓冲流的使用

通过实验掌握缓冲（字符）流 BufferedReader 类和 BufferedWriter 类的使用。使用 Eclipse 创建 Java 项目"task9_5"，在该项目中创建一个名为"Task9_5"的 Java 主

类。Task9_5.java文件中的代码如图9-6所示(其中包含了CodeLineCounter类的代码)。

```
//代码行数统计类
class CodeLineCounter{
    public int totalCount=0;                    //代码行数计数器
    //方法countLines用于统计文件fileName的行数
    public boolean countLines(String fileName){
        try{
            int temp = 0;
            FileReader reader = new FileReader(fileName);     //根据文件名创建文件字符输入流reader
            BufferedReader bufferedReader = ▇▇▇;              //根据文件字符输入流对reader创建缓冲输入流
            String curLine = bufferedReader.▇▇▇;              //从缓冲输入流中读取一行内容
            while(▇▇▇){                                        //若读取的字符串不是null
                temp++;                                         //计数
                curLine = bufferedReader.readLine();            //从缓冲输入流中读取下一行内容
            }
            bufferedReader.close();                             //关闭缓冲流
            totalCount = totalCount + temp;
            return true;
        }
        catch(Exception e){
            System.out.println(e.getMessage());
            return false;
        }
    }
    //方法deleteSpcLine用于删除文件sFile中的空行,并把删除空行后的内容写入到文件dFile中
    public boolean deleteSpcLine(String sFile,String dFile){
        try{
            File inFile = new File(sFile);
            FileReader reader = ▇▇▇;                           //根据文件对象inFile创建文件字符输入流
            BufferedReader bufferedReader = ▇▇▇;               //根据文件字符输入流对象reader创建缓冲输入流
            File outFile = new File(dFile);
            FileWriter writer = new FileWriter(outFile);        //根据文件对象outFile创建文件字符输出流
            BufferedWriter bufferedWriter = new BufferedWriter(writer);  //根据文件字符输出流对writer创建缓冲输出流
            boolean spcFlag = false;                            //用于标记前一行是否是空行
            String curLine = bufferedReader.readLine();         //从缓冲输入流中读取一行内容
            while(curLine!=null){                               //若读取的字符串不是null
                if(▇▇▇){                                        //若字符串对象curLine等于空字符串
                    spcFlag = true;                             //标记当前行是空行
                }
                else {
                    if(spcFlag)▇▇▇                              //若前一行是空行,则向缓冲输出流中写入一个新行
                    bufferedWriter.write(curLine);              //则向缓冲输出流中写入当前行的内容
                    bufferedWriter.newLine();                   //向缓冲输出流中写入一个新行
                    spcFlag = false;                            //标记当前行不是空行
                }
                curLine = ▇▇▇;                                  //从缓冲输入流中读取下一行内容
            }
            bufferedReader.close();
            bufferedWriter.close();                             //关闭缓冲流
            return true;
        }
        catch(Exception e){
            System.out.println(e.getMessage());
            return false;
        }
    }
}
public class Task9_5 {
    public static void main(String[] args) {
        CodeLineCounter cLCounter = new CodeLineCounter();
        cLCounter.deleteSpcLine("d:\\MyCalendar.java","d:\\MyCalendar_New.java");  //换成相应的文件名
        cLCounter.countLines("d:\\MyCalendar_New.java");
        System.out.println("共有代码"+cLCounter.totalCount+"行");
    }
}
```

图9-6 Task9_5.java文件代码

将图9-6所示代码中被涂黑部分补充完整,从而实现删除文件中空行(单个空行不删除,若有多个连续的空行则仅保留一个空行)、统计文件中行数的功能。

（六）对象流的使用

通过实验掌握对象流 ObjectOutputStream 类和 ObjectInputStream 类的使用。使用 Eclipse 创建 Java 项目"task9_6"，在该项目中创建一个名为"Task9_6"的 Java 主类。Task9_6.java 文件中的代码如图 9-7 所示（其中包含了 Student 类的代码）。

```java
class Student implements Serializable{    //需要在对象输入输出流中操作的类必须是可序列化的
    private String sID;
    private String name;
    int grade=1;
    int classNum=1;
    public Student(String sID, String name) {
        super();
        this.sID = sID;
        this.name = name;
    }
    public String getsID() {
        return sID;
    }
    public String getName() {
        return name;
    }
}
public class Task9_6 {
    public static void main(String[] args) {
        try{
            Student student = new Student("971115", "yly");
            student.grade = 1997;
            student.classNum = 1;
            //写入操作
            FileOutputStream out = new FileOutputStream("d:\\object.dat");  //创建文件字节输出流对象out
            ObjectOutputStream outObject = ■■■■■■■;  //使用字节输出流对象out创建对象输出流
            outObject.■■■■■;    //把student对象写入到对象输出流中
            outObject.close();    //关闭对象输出流
            //读取操作
            FileInputStream in = new ■■■■■■■;    //创建文件字节输入流对象
            ObjectInputStream inObject = new ObjectInputStream(in);  //使用字节输入流对象in创建对象输入流
            student = (Student)■■■■■;    //从对象输入流读取对象，需要强制类型转换
            inObject.close();    //关闭对象输入流
            //输出student对象各成员变量的值
            System.out.println(student.getsID());
            System.out.println(student.getName());
            System.out.println(student.grade);
            System.out.println(student.classNum);
            System.out.println("OK");
        }
        catch(Exception e){
            System.out.println(e.getMessage());
        }
    }
}
```

图 9-7 Task9_6.java 文件代码

将图 9-7 所示代码中被涂黑部分补充完整，从而实现向"d:\object.dat"文件中写入 student 对象，并重写从"d:\object.dat"文件中读取 student 对象的功能。

（七）使用标准数据流的应用程序

标准数据流指在字符方式下（如 DOS 提示符）程序与系统进行输入/输出的方式。键盘和显示器屏幕是标准输入、输出设备，数据输入的起点为键盘，数据输出的终点是屏幕，输出的数据可以在屏幕上显示出来。

1. 程序功能

将键盘上输入的字符在屏幕上显示出来。

2. 编写 KY9_7.java 程序文件

源代码如下。

```java
class KY9_7{
    public static void main(String[] args) throws java.io.IOException {
        byte buffer[]=new byte[10];
        System.out.println("从键盘输入不超过 10 个字符,按 Enter 键结束输入:");
        int count = System.in.read(buffer);//读取输入的字符并存放在缓冲区 buffer 中
        System.out.println("保存在缓冲区 buffer 中元素的个数为"+count);
        System.out.println("buffer 中各元素的值为");
        for(int i=0;i<count;i++){
            System.out.print(" "+ buffer[i]);//在屏幕上显示 buffer 元素的值
        }
        System.out.println();
        System.out.println("输出 buffer 字符元素:");
        System.out.write(buffer, 0, buffer.length);
    }
}
```

3. 编译并运行该文件

编译并运行 KY9_7.java 文件。

(八) 随机读写流的使用

掌握随机读写流 RandomAccessFile 类的使用。使用 Eclipse 创建 Java 项目"task9_8",在该项目中创建一个名为"Task9_8"的 Java 主类。Task9_8.java 文件中的代码如图 9-8 所示。

将图 9-8 所示代码中被涂黑部分补充完整,从而实现向文件"d:\data.dat"中写入 10 个 10~99 之间的随机数,并对写入文件中的 10 个整数采用选择法进行排序,最后输出文件中排序后的 10 个整数的功能。

```
public class Task9_8 {
    public static void main(String[] args) {
        try{//创建随机读写流对象,同时支持读和写
            RandomAccessFile randomFile = new RandomAccessFile("d:\\data.dat",    );
            for(int i=0;i<10;i++){                         //向文件中写入10个随机数
                int n = (int)(Math.random()*90+10);
                randomFile.        ;                        //向随机读写流中以int型写入整数n
            }
            for(int i=0;i<9;i++){                          //选择排序
                randomFile.        ;                        //定位到第i个整数的位置(每个int型数据占4个字节)
                int iData = randomFile.readInt();           //读取一个整数到变量iData
                for(int j=i+1;j<10;j++){
                    randomFile.seek(j*4);                   //为了读取第j个整数,需要定位到第4*j个位置
                    int jData = randomFile.        ;         //读取一个整数到变量jData
                    if(jData>iData){
                        randomFile.seek(i*4);               //定位到第i个整数的位置
                        randomFile.writeInt(jData);         //把jData写入到第i个整数的位置
                        randomFile.seek(j*4);               //定位到第j个整数的位置
                        randomFile.writeInt(iData);         //把jData写入到第j个整数的位置
                        iData = jData;                      //让iData记录较大值
                    }
                }
            }
            for(int i=0;i<10;i++){                          //从文件中读出排序后的值
                randomFile.seek(i*4);                       //定位到第i个整数的位置
                int iData = randomFile.readInt();
                System.out.println(iData);
            }
            randomFile.close();
        }
        catch(Exception e){;}
    }
}
```

图9-8 Task9_8.java文件代码

(九)简单的文本编辑器

1. 程序功能

创建一个简单的文本编辑器,可打开文件对话框选择打开一个文件,并在文本区进行编辑,然后把它保存起来。

2. 编写LX9_9.java程序文件

源代码如下。

```
import java.io.*;
import java.awt.*;
import java.awt.event.*;
public class LX9_9 extends Frame implements ActionListener {
    FileDialog fileDlg;
    String str,fileName;
    byte byteBuf[]=new byte[10000];
```

```
TextArea ta=new TextArea();
MenuBar mb=new MenuBar();
Menu m1=new Menu("文件");
MenuItem open=new MenuItem("打开");
MenuItem close=new MenuItem("关闭");
MenuItem save=new MenuItem("保存");
MenuItem exit=new MenuItem("退出");
LX9_9() {
  setTitle("简易文本编辑器");
  setSize(400,280);
  add("Center", ta);
  addWindowListener(new WindowAdapter() {
    public void windowClosing(WindowEvent e) {
      System.exit(0);
    }
  }
  m1.add(open);
  m1.add(close);
  m1.add(save);
  m1.addSeparator();
  m1.add(exit);
  open.addActionListener(this);
  close.addActionListener(this);
  save.addActionListener(this);
  exit.addActionListener(this);
  mb.add(m1);
  setMenuBar(mb);
  show();
}
public void actionPerformed(ActionEvent e) {
  if (e.getSource()==exit)
    System.exit(0);
```

```java
        else if (e.getSource()==close) //关闭文件
            ta.setText(null); //设置文本区为空
        else if (e.getSource()==open) { //打开文件
            fileDlg=new FileDialog(this,"打开文件"); // 生成文件对话框
            fileDlg.show(); //显示文件对话框
            fileName=fileDlg.getFile(); //获取文件名
            try {
                FileInputStream in=new FileInputStream(fileName); //建立文件输入流
                in.read(byteBuf); //将文件内容读到字节数组
                in.close(); //关闭文件输入流
                str=new String(byteBuf); //将字节数组转换成字符串
                ta.setText(str); //将字符串显示在文字区
                setTitle("简易文本编辑器 — "+fileName);
            } catch(IOException ioe) {}
        }
        else if (e.getSource()==save) { //保存文件
            fileDlg=new FileDialog(this,"保存文件",FileDialog.SAVE); // 生成文件对话框
            fileDlg.show();
            fileName=fileDlg.getFile();
            str=ta.getText(); //将文本区内容读至字符串
            byteBuf=str.getBytes(); //将字符串转换成字节数组
            try {
                FileOutputStream out=new FileOutputStream(fileName); //建立文件输出流
                out.write(byteBuf); //将字节数组写入文件输出流
                out.close(); //关闭文件输出流
            } catch(IOException ioe) {}
        }
    }
    public static void main(String args[]) {
        new LX9_9();
    }
}
```

3. 编译并运行程序

编译并运行程序,会出现如图9-9所示窗口。

图9-9 简易文本编辑器窗口

4. 弹出文件对话框

单击"文件"下拉菜单中的"打开"菜单项,将弹出如图9-10所示文件对话框。

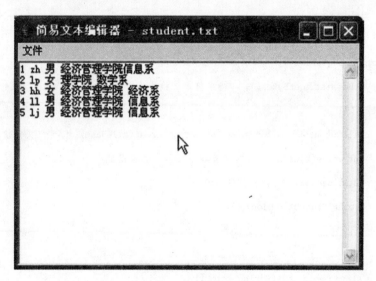

图9-10 文件对话框

5. 选取打开文本操作

从文件列表中选取要打开的文本文件,则文件的内容会显示在文本编辑区,如图9-11所示。在文本编辑区中,可以进行常规的编辑操作。尽管没有添加弹出式菜单,但右击时还会有一个系统的弹出式编辑菜单出现。

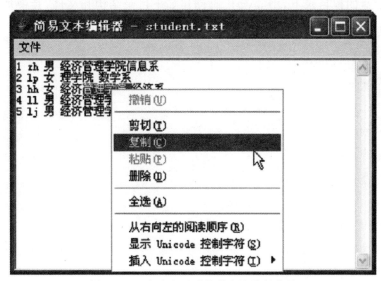

图9-11　右击出现系统弹出式编辑菜单

6. 保存文件操作

选取"文件"下拉菜单中的"保存"菜单项将打开保存文件对话框,如图9-12所示。可直接输入文件名,也可从文件显示区选择一个文件名,此时将弹出一个警告对话框,让你确认是否覆盖原文件。

图9-12　保存文件对话框

（十）综合实验

1. 创建目录结构和文件

在 C：盘的根目录下，用 java 代码创建如下目录结构和文件，其中加粗标记的为文件，其他为盘符或文件夹。

C：\
    MyData.txt
    成绩
        Java 程序设计
        文学欣赏
        数学分析
    资料
        项目申报
        国家自然科学基金.doc
        一等奖学金申报.txt
    杂学
        饮食与文化的关系.pdf
        如何让自己更美.pdf

2. 读写文本文件

编程实现在 C 盘创建一个文本文件"MyData.txt"，并向该文本文件中写入内容："我是中国人，我热爱中国；我是师大人，我为师大添光彩"；再从文本文件中读取其中的内容，显示到控制台上（使用字节流实现）。

四、思考题

（1）Java 的 I/O 流按照处理时的数据单位可以分成哪些类？

（2）编写一个程序，读入一个文本文件，显示文件中包含的字符数和行数。

实验十 线程

一、实验目的

- 线程的概念、线程的生命周期。
- 多线程的编程：继承 Thread 类与使用 Runnable 接口。
- 使用多线程机制实现动画。

二、实验要求

- 掌握利用 Java 语言编写多线程程序的方法。
- 掌握线程的调度方法。
- 掌握多线程环境中 GUI 程序的编写方法。

三、实验内容

（一）Thread 子类的方法实现多线程

1. 编写 KY10_1.java 程序文件

源代码如下。

```
class SimpleThread extends Thread {
    public SimpleThread(String str) {
        super(str);
    }
    public void run() {
```

```
        for (int i = 0; i < 10; i++) {
            System.out.println(i + " " + getName());
            try {
                sleep((int)(Math.random() * 1000));
            } catch (InterruptedException e) {}
        }
        System.out.println("DONE! " + getName());
    }
}
public class TwoThreadsTest {
    public static void main (String[] args) {
        new SimpleThread("Go to Beijing?? ").start();
        new SimpleThread("Stay here!! ").start();
    }
}
```

2. 编译并运行该程序

编译并运行 KY10_1.java 程序。

（二）通过实验掌握通过 Runnable 接口类实现多线程

使用 Eclipse 创建 Java 项目"task10_2"，在该项目中创建一个名为"Task10_2"的 Java 主类。Task10_2.java 文件中的代码如图 10-1 所示（其中包含 ObjectRunnable 类的代码）。

```
//声明一个实现Runnable接口的类，并实现接口中定义的run方法
class ObjectRunnable implements Runnable{
    public void run() {
        for(int i=0;i<1000;i++){
            System.out.println("Wang!Wang!");
        }
    }
}
//主线程（单线程的例子）
public class Task10_2 {
    public static void main(String[] args) {
        FirstThread myFirstThread = new FirstThread();
        myFirstThread.start();

        Thread mySecondThread = new Thread(new ObjectRunnable());
        mySecondThread.start();

        for(int i=0;i<1000;i++){
            System.out.println("hello");
        }
    }
}
```

图 10-1　Task10_2.java 文件代码

(三) 利用 Runnable 接口的方法实现多线程

1. 程序功能

一个时钟 Applet,能显示当前时间并逐秒进行更新。

2. 编写 KY10_2.java 程序文件

源代码如下。

```java
import java.awt.*;
import java.applet.*;
import java.util.*;
public class Clock extends Applet implements Runnable{
  Thread clockThread;
   public void start(){
     if(clockThread==null){
       clockThread=new Thread(this,"Clock ");
       clockThread.start();
     }
   }
   public void run(){
     while(clockThread！=null){
       repaint();
       try{
         clockThread.sleep(1000);
       }catch(InterruptedException e){}
     }
   }
   public void paint(Graphics g){
     Date now=new Date();
     g.drawString(now.getHours()+";"+now.getMinutes()+";"+now.getSeconds(),5,10);
   }
   public void stop(){
     clockThread.stop();
```

```
        clockThread=null;
    }
}
```

3. 编译并运行该程序

编译并运行 KY10_2.java 程序。

四、思考题

（1）简述并区分程序、进程和线程三个概念。

（2）线程有哪几个基本的状态？Java 中线程调度遵循何种原则？

（3）实现多线程可以用哪两种基本方法？将这两种方法进行比较。

实验十一　网络编程

一、实验目的

- 掌握 InetAddress 类的使用。
- 掌握 TCP 与 UDP 编程：Socket 与 Datagram 的概念和编程方法。
- 掌握 URL 类的使用：URL 的概念和编程。

二、实验要求

- 通过 Socket 编程，掌握网络应用程序的开发方法。
- 完成数据库的连接。
- 掌握利用 Java 提供的基本组件进行网络传输的方法。
- 掌握 Java 提供的多线程机制、异常处理机制和低层对协议的通信机制，通过 Socket 编程，掌握网络应用程序的开发方法。
- 设计测试，性能评估。上机练习之前，必须先完成程序的书写，再上机调试。

三、实验内容

（一）使用 InetAddress 类的方法

通过使用 InetAddress 类的方法，获取 http://www.ytu.edu.cn/ 的主机的 IP 地址；获取本地机的名称和 IP 地址。

(二) Socket 编程

使用 Socket 编程，客户机发送数据到服务器，服务器将接收到的数据返回给客户机。

(三) UDP 编程

使用 UDP 编程，客户机发送数据到服务器，服务器将接收到的数据返回给客户机。

(四) 获取 URL 信息

1. 编写 KY11_1.java 程序文件

源代码如下。

```java
import java.net.*;
import java.io.*;
public class URLTest {
    public static void main(String[] args){
        URL url=null;
        InputStream is;
        try{
            url=new URL("http://localhost/index.html");
            is=url.openStream();
            int c;
            try{
                while((c=is.read())!=-1)
                    System.out.print((char)c);
            }catch(IOException e){
            }finally{
                is.close();
            }
        }catch(MalformedURLException e){
            e.printStackTrace();
        }catch(IOException e){
            e.printStackTrace();
        }
```

```java
        System.out.println("文件名:"+url.getFile());
        System.out.println("主机名:"+url.getHost());
        System.out.println("端口号:"+url.getPort());
        System.out.println("协议名:"+url.getProtocol());
    }
}
```

2. 编译并运行程序

编译并运行以上代码。

(五) 利用 URL 类获取网络资源

1. 编写 KY11_2.java 程序文件

源代码如下。

```java
import java.net.*;
import java.io.*;
public class URLReader {
    public static void main(String[] args) throws Exception {
        URL web = new URL("http://166.111.7.250:2222/");
        BufferedReader in = new BufferedReader(new InputStreamReader(web.openStream()));
        String inputLine;
        while ((inputLine = in.readLine()) != null) System.out.println(inputLine);
        in.close();
    }
}
```

2. 编译并运行程序

编译并运行以上代码。

(六) 利用 URLConnection 读取 URL 资源

1. 编写 KY11_3.java 程序文件

源代码如下。

```java
import java.net.*;
import java.io.*;
public class URLConnectionReader {
    public static void main(String[] args) throws Exception {
```

```
        URL web = new URL("http://166.111.7.250:2222/");
//get an instance of URLConnection
        URLConnection webc=web.openConnection();
//use of URLConnection
        BufferedReader in = new BufferedReader(new InputStreamReader(
                 webc.getInputStream()));
        String inputLine;
        while ((inputLine = in.readLine()) != null) System.out.println(inputLine);
        in.close();
    }
}
```

2. 编译并运行程序

编译并运行以上代码。

（七）掌握 URLConnection 对 URL 资源的写入

1. 编写 KY11_4.java 程序文件

源代码如下。

```
import java.io.*;
import java.net.*;
public class Reverse {
    public static void main(String[] args) throws Exception {
      if (args.length != 1) {
        System.err.println("Usage:  java Reverse string_to_reverse");
        System.exit(1);
      }
      String stringToReverse=args[0];
      URL url = new URL("http://java.sun.com/cgi-bin/backwards");
      URLConnection connection = url.openConnection();
      connection.setDoOutput(true);
      PrintWriter out = new PrintWriter(connection.getOutputStream());
      out.println("string=" + stringToReverse);
      out.close();
```

```
    BufferedReader in = new BufferedReader(new InputStreamReader(
                                   connection.getInputStream()));
    String inputLine;
    while ((inputLine = in.readLine()) ! = null) System.out.println(inputLine);
      in.close();
    }
}
```

2. 编译并运行

(八) 网络编程综合实验

1. 设计一个程序获取本机 IP

判断该 IP 的类别,并区别是 IPv4 还是 IPv6。

2. 编写程序

要求利用 URL,读取 URL 文件并显示出来。

3. 实现数据的通信与接收

根据下面的实验步骤,应用 UDP 通信方式,实现数据的发送与接收。

4. 应用 Socket 通信方式,编写 C/S 程序,要求客户端发送圆的半径,服务器端接收后,计算出圆的面积,并将结果返回给客户端

服务器端代码如下。

```
try {
    ServerSocket serverSocket = new ServerSocket(8000);
    Socket connectToClient = serverSocket.accept();
    DataInputStream isFromClient = new DataInputStream(connectToClient.getInputStream());
    DataOutputStream osToClient = new DataOutputStream(connectToClient.getOutputStream());
    while(true) {
        double radius = isFromClient.readDouble();
        if (radius == -1.0) {
            serverSocket.close();
            return;
        }
        System.out.println("radius from client: " + radius);
```

```
            double area = radius * radius * Math.PI;
            osToClient.writeDouble(area);
            osToClient.flush();
            System.out.println("Area is: " + area);
        }
    }
    catch(IOException e){
        System.err.println(e);
    }
}
```

客户端代码如下。

```
try {
    Socket connectToServer = new Socket("localhost", 8000);
    DataInputStream isFromServer = new DataInputStream(connectToServer.getInputStream());
    DataOutputStream osToServer = new DataOutputStream(connectToServer.getOutputStream());
    while(true) {
        System.out.println("Please enter a radius: ");
        BufferedReader br = new BufferedReader(new InputStreamReader(System.in));
        double radius = Double.parseDouble(br.readLine());
        osToServer.writeDouble(radius);
        osToServer.flush();
        double area = isFromServer.readDouble();
        System.out.println("Area from the server is: " + area);
    }
}
catch(IOException e){
    System.err.println();
}
```

5. 运行程序

运行以下程序,分析此端口扫描程序,并给出程序运行结果。

```
String host = "www.hebtu.edu.cn";
for (int i=1; i<512; i++) {
```

```
    try {
        Socket s = new Socket(host, i);
        System.out.println("主机" + host + "监听端口" + i);
    }
    catch(UnknownHostException e) {
        System.out.println(e);
    }
    catch(IOException e) {
    }
}
```

6. 此实验的具体实现步骤

(1) 实验题目(1)实验步骤。

①在项目 Experiment11 中新建类 Question11_5,生成文件 Question11_5.java。

②在文件 Question11_5.java 中添加如下所示包声明和引用包语句。

```
package experiment11;
import java.net.*;
```

③在类 Question11_5 中添加方法 run(),并编写如下代码。

```
try {
    InetAddress inetAdd = InetAddress.getLocalHost();
    byte[] address = inetAdd.getAddress();
    if (address.length == 4) {
        System.out.println(" The IP version is IPv4 ");
        int firstbyte = address[0];
        if (firstbyte < 0)
            firstbyte += 256;
        if ((firstbyte & 0x80) == 0)
            System.out.println(" The IP class is A ");
        else if ((firstbyte & 0xC0) == 0x80)
            System.out.println(" The IP class is B ");
        else if ((firstbyte & 0xE0) == 0xC0)
```

```
                System.out.println("The IP class is C");
            else if ((firstbyte & 0xF0) == 0xE0)
                System.out.println("The IP class is D");
            else if ((firstbyte & 0xF8) == 0xF0)
                System.out.println("The IP class is E");
        }
        else if (address.length == 16)
            System.out.println("The IP version is IPv6");
    }
    catch(Exception e) {
        System.out.println(e.getMessage());
    }
```

④在文件 Main.java 中,添加如下代码。

```
Question11_5 questionInstance11_5 = new Question11_5();
questionInstance11_5.run();
```

⑤按 F6 键运行程序。

注意:结果会因网络环境的不同而不同。

(2) 实验题目(2)实验步骤。

①在项目 Experiment11 中新建类 Question11_6,生成文件 Question11_6.java。

②在文件 Question11_6.java 中添加如下所示包声明和引用包语句。

```
package experiment11;
import java.net.*;
import java.io.*;
```

③在类 Question11_6 中添加方法 run(),并编写如下代码。

```
InputStream fileConn = null;
String fileLine;
String url = "http://www.hebtu.edu.cn";
URL fileURL;
try {
    fileURL = new URL(url);
```

```
        fileConn = fileURL.openStream();
        BufferedReader br = new BufferedReader(new InputStreamReader(fileConn));
        while ((fileLine = br.readLine()) != null)
            System.out.println(fileLine + "\n");
    }
    catch (IOException e) {
        System.out.println("Error in I/O: " + e.getMessage());
    }
```

④在文件 Main.java 中,添加如下代码。

```
Question11_6 questionInstance11_6 = new Question11_6();
questionInstance11_6.run();
```

⑤按 F6 键运行程序。

注意:结果会因服务器端网页内容的不同而不同。

(3) 实验题目(3)实验步骤。

①在项目 Experiment11 中新建类 Question11_7,生成文件 Question11_7.java。

②在文件 Question11_7.java 中添加如下所示包声明和引用包语句。

```
package experiment11;
import java.net.*;
import java.io.*;
import java.util.Scanner;
```

③添加 UDPServer 类,声明其私有变量。

```
private DatagramSocket server;
private DatagramPacket packet;
private byte bReceive[];
private String strRecive;
```

④添加 UDPServer 类构造函数 UDPServer()。

```
try {
    server = new DatagramSocket(10005);
    bReceive = new byte[1024];
    packet = new DatagramPacket(bReceive, bReceive.length);
```

```
System.out.println("服务器端准备好,正在等待接收数据");
server.receive(packet);
strRecive = new String(bReceive,0,packet.getLength());
System.out.println("接收数据成功！\n数据为" + strRecive);
}
catch(Exception e) {
    e.printStackTrace();
}
```

⑤添加 UDPClient 类,声明其私有变量。

```
private DatagramSocket client;
private DatagramPacket packet;
private byte bSend[];
private String strSend;
```

⑥添加 UDPClient 类构造函数 UDPClient()。

```
try {
    client = new DatagramSocket(10002);
    bSend = new byte[1024];
    System.out.println("请输入发送数据：");
    Scanner cin = new Scanner(System.in);
    strSend = cin.next();
    bSend = strSend.getBytes();
    packet = new DatagramPacket(bSend, bSend.length, InetAddress.getByName("localhost"),10005);
    client.send(packet);
}
catch(SocketException se) {
    se.printStackTrace();
}
catch(IOException ie) {
    ie.printStackTrace();
}
```

⑦在类 Question11_7 中添加方法 run()，并编写如下代码。

```
int flag = 1;//操作标志
System.out.print("请输入操作类型：\r\n1（服务器端）\r\n2（客户端）");
Scanner cin = new Scanner(System.in);
flag = cin.nextInt();
switch(flag){
    case 1:
        new UDPServer();
        break;
    case 2:
        new UDPClient();
        break;
    default:
        System.out.println("您输入的操作为非法操作！");
}
```

⑧在文件 Main.java 中，添加如下代码。

```
Question11_7 questionInstance11_7 = new Question11_7();
questionInstance11_7.run();
```

⑨按 F6 键运行程序，在 Output－Experiment11（run）中，首先选择 1 启动"服务器"。

⑩再按 F6 键运行程序，在 Output－Experiment11（run）♯2 中，选择 2 启动"客户端"。之后输入要发送的内容，如"Hello"，按 Enter 键。此时，服务器端应该显示出效果。

四、思考题

（1）什么是 URL？一个 URL 地址由哪些部分组成？

（2）网络环境下的 C/S 模式的基本思想是什么？什么是客户机？什么是服务器？它们各自的作用如何？C/S 模式的基本工作过程如何？

（3）简述流式 Socket 的通信机制及它的最大特点。

（4）数据报（UDP）通信有何特点？简述 Java 实现数据报通信的基本工作过程。

实验十二 数据库的连接：JDBC

一、实验目的

了解 JDBC 核心 API，利用 JDBC 核心 API，建立数据库连接、执行 SQL 语句、取得查询集、数据类型支持等功能。

二、实验要求

- 了解 JDBC 的概念和工作原理。
- 掌握使用 JDBC 实现简单的数据库管理。

三、实验内容

（一）配置 ODBC 数据源

配置 ODBS 数据源主要步骤如下。

(1) 从开始菜单中，选择"设置"—"控制面板"。

(2) 在控制面板中选择"32 位 ODBC"。

(3) 打开"32 位 ODBC"后，看到的应该是一个卡片式对话框，上面一排有多个卡片标签，其中包括"用户 DSN""系统 DSN""文件 DSN"等。选择"系统 DSN"选项。

(4) 单击"系统 DSN"中的"添加"按钮，弹出一个对话框。

(5) 在对话框中,列出了当前系统支持的 ODBC 数据源的驱动程序,选择"Microsoft Access Driver",单击"完成"按钮,弹出一个对话框。

(6) 在对话框中,向"数据源"文本框内输入数据源的名字,这个名字可以任取,在这个例子中,我们输入的名字是"vfox"。然后,单击"创建"按钮。

(7) 在对话框中,选择数据库存放的目录,然后向"数据库名"文本框内输入数据库的名字,这个名字可以任取,在这个例子中,我们输入的名字是"vfox"。然后,单击"确定"按钮,会弹出"数据库创建成功"的提示对话框。

(8) 单击"确定"按钮。

(二) 编写程序

建立"Student"表可按照表 12-1 的结构建立"student"表。

表 12-1 "student"表内容

字段名	Java 数据类型	宽度	SQL 数据类型
Name	String	10	Char(10)
Sex	String	2	Char(2)
Age	Int	3	Integer

(三) 编写程序,完成填写功能

向"student"表中填入若干数据记录。

(四) 编写程序,完成查询功能

在"student"表中分别查询所有记录以及满足条件"age>18"的记录。

四、思考题

(1) 什么是 SQL 语言?与数据库前端操作有关的 SQL 语句主要有哪些?它们的功能如何?

(2) JDBC 的主要功能是什么?它由哪些部分组成?JDBC 中驱动程序的主要功能是什么?

(3) 简述 Java 程序中使用 JDBC 完成数据库操作的基本步骤。

(4) 什么是数据库连接?为什么在做数据库操作之前要首先完成数据库的连接?Java 中如何实现与后台数据库的连接?

实验十三　JSP 与 Servlet 技术

一、实验目的

- 理解 JSP 元素的概念。
- 理解 JSP 页面中生成静态内容和动态内容的机制。
- 理解 JSP 页面的服务请求通过 Servlet 的执行机制。

二、实验要求

- 掌握 Servlet 的实现方法。
- 掌握 JSP 页面的创建。

三、实验内容

（一）Java Web Server 开发环境的配置

1. 程序安装

JWS1.1 开发环境安装在"C：\JavaWebServer1.1\"目录下。

2. 设置环境变量 CLASSPATH

如果在 autoexec.bat 中没有进行设置，则进入命令行（MS-DOS）方式，进行如下设置：SET　CLASSPATH＝C：\ JavaWebServer1.1 \ Lib \ jws.jar；%CLASSPATH%。

3. 启动 JavaWebServer

进入命令行(MS－DOS)方式,将当前目录设置为"C：\JavaWebServer1.1\Bin",运行 httpd.exe。C：\JavaWebServer1.1\Bin＞httpd.exe。

4. 显示缺省主页

在浏览器上输入 URL"http：//localhost：8080/"。

（二）Java Servlet 程序开发过程

1. 进入 MS－DOS

进入命令行(MS－DOS)方式。

2. 设置环境变量 CLASSPATH

如果在"C：\autoexec.bat"中没有进行设置,则命令行指令如下所示：SET CLASSPATH＝C：\JavaWebServer1.1\Lib\jws.jar；％CLASSPATH％。

3. 启动 Java Web Server

命令行指令如下所示：C：\JavaWebServer1.1\Bin＞httpd.exe。

4. 键入程序

在编辑软件中键入下面的程序,文件名为"SimpleServlet.java",将该文件保存在"D：\Java\"目录中。

```java
import java.io.*;
import javax.servlet.*;
import javax.servlet.http.*;
public class SimpleServlet extends HttpServlet{
    int connections;        //用于记数
    public void init(ServletConfig conf) throws   ServletException{
        super.init(conf);
        nections=0;
    }
    //获得一个浏览器连接的链路,用于发送输出结果
    public void service(HttpServletRequest req, HttpServletResponseres)  throws ServletException,IOException{
        ServletOutputStream  out=res.getOutputStream();
        res.setContentType("text/html");//设置应答内容的 MIME 类型
```

```
            out.println("<HTML><HEAD><TITLE>Servlets</TITLE></HEAD>");
            out.println("<body>Say hello to Java Servlet  Programming,");
            String str=req.getParameter("userName");
            if(str!=null){
                out.println(str);
                out.println("<p>number: ");
                connections++;
                out.println(Integer.toString(connections));
                out.println("</body></html>");
                out.close();   //关闭输出流
            }
        }
    }
```

5. 进入命令行(MS-DOS)方式,运行Java编译器

命令行指令如下所示。

D:\Java>javac SimpleServlet.java。

6. 保存文件

将生成的class文件保存在"C:\JavaWebServer1.1\servlets\"目录中。

7. 运行

(1) 在浏览器中输入下面的URL地址。

http://localhost:8080/servlet/SimpleServlet? userName=aaa。

(2) 在浏览器中可以看到输出的结果如下。

  Say hello to Java Servlet Programming, aaa

      number: 1

(3) 如果再次访问该Servlet,返回的结果中的第二句会变成如下所示。

      number: 2

(三) Java Server Web Development Kit 1.0.1 开发环境

1. 开发环境安装

JSWDK1.0.1开发环境安装在"C:\jswdk-1.0.1\"目录下。

2. 设置环境变量

如果在"C:\jswdk-1.0.1\startserver.bat"中没有进行设置,则进入命令行

(MS-DOS)方式,进行如下设置:SET JAVA_HOME=C:\JDK1.2.1。目的是确定 startserver.bat 中的 set sysJars=%JAVA_HOME%\lib\tools.jar。否则,JSP 页面执行时,可能会出现错误"Error:500sun/tools/javac/Main",这是因为找不到类 sun/tools/javac/Main。

3. DOS 窗口的属性设置

打开当前 DOS 窗口的属性窗口,选择"内存"选项卡,把"初始环境"旁边的下拉式列表从"自动"改成一个大于等于 2816 的数字。否则,在执行 startserver.bat 时,在 MS-DOS 窗口中可能出现错误"Out of environment space",这是因为 Windows 给环境变量分配的空间太小了。

4. 启动 Java Server Web Development Kit

进入命令行(MS-DOS)方式,将当前目录设为"C:\jswdk-1.0.1",然后运行 startserver.bat:C:\jswdk-1.0.1> startserver.bat。

5. 显示缺省主页

在浏览器上输入 URL"http://localhost:8080/"。

6. 停止 Java Server Web Development Kit

进入命令行(MS-DOS)方式,将当前目录设为"C:\jswdk-1.0.1",然后运行 stopserver.bat:C:\jswdk-1.0.1>stopserver.bat。

(四)掌握 JSP 页面开发过程

1. 进入 MS-DOS 方式

进入命令行(MS-DOS)方式

2. 设置环境变量

在"C:\jswdk-1.0.1\startserver.bat"中没有进行设置时进行此处的环境变量设置。

SET JAVA_HOME=C:\JDK1.2.1

3. MS-DOS 窗口的属性设置

打开当前 MS-DOS 窗口的属性窗口,选择"内存"选项卡,把"初始环境"旁边的下拉式列表从"自动"改成一个大于等于 2816 的数字。

4. 启动 Java Server Web Development Kit

C:\jswdk-1.0.1>startserver.bat

5. 使用编辑软件键入下面的程序

文件名为"temp.jsp",将其保存在"C:\jswdk-1.0.1\example\jsp\"目录中。

```
<HTML>
<HEAD>
<TITLE>JSP Date Demo Page</TITLE>
</HEAD>
<BODY>
<H1>JSP Date Demo Page</H1>
The current date is
<%
java.util.Date date = new java.util.Date();
out.println(date);
%>.
<br>expression syle
<%=date%>
</BODY>
</HTML>
```

6. 运行

在浏览器中输入 URL 地址:http"//localhost:8080/example/jsp/temp.jsp",在浏览器中可以看到输出的结果如下。

JSP Date Demo Page

The current date is Sat Nov 18 03:01:06 CST 2000 .

expression syle Sat Nov 18 03:01:06 CST 2000

(五) 开发 Java Servlet 程序

使用 Java Server Web Development Kit 1.0.1 开发环境开发 Java Servlet 程序。

1. 进入 MS-DOS 方式

进入命令行(MS-DOS)方式。

2. 设置环境变量 JAVA_HOME

在"C:\jswdk-1.0.1\startserver.bat"中没有进行设置时进行此处的环境变

量设置,命令行指令如下所示。

SET JAVA_HOME=C:\JDK1.2.1

3. 设置环境变量 CLASSPATH

命令行指令如下所示。

SET CLASSPATH=c:\jswdk-1.0.1\lib\servlet.jar;%CLASSPATH%

4. MS-DOS 窗口的属性设置

打开当前 MS-DOS 窗口的属性窗口,选择"内存"选项卡,把"初始环境"旁边的下拉式列表从"自动"改成一个大于等于 2816 的数字。

5. 启动 Java Server Web Development Kit

命令行指令如下所示。

C:\jswdk-1.0.1>startserver.bat。

6. 在编辑软件中键入下面的程序

文件名为"SimpleServlet.java",将该文件保存在"D:\Java"目录中。

```java
import java.io.*;
import javax.servlet.*;
import javax.servlet.http.*;
    public class SimpleServlet extends HttpServlet{
      int connections;         //用于记数
        public void init(ServletConfig conf) throws
              ServletException{
        super.init(conf);
        connections=0;
        }
        public void service(HttpServletRequest req, HttpServletResponse res)
                throws  ServletException,IOException{
        //获得一个浏览器连接的链路,用于发送输出结果
        ServletOutputStream
              out=res.getOutputStream();
        res.setContentType("text/html");//设置应答内容的 MIME 类型
        out.println("<HTML><HEAD><TITLE>Servlets</TITLE></HEAD>");
        out.println("<body>Say hello to Java Servlet   Programming,");
```

```
        String str=req.getParameter("userName");
        if(str!=null)
        out.println(str);
        out.println("<p>number: ");
        connections++;
        out.println(Integer.toString(connections));
        out.println("</body></html>");
        out.close();  //关闭输出流
    }
}
```

7. 进入命令行(MS-DOS)方式,运行 Java 编译器

命令行指令如下所示。

D:\Java>javac SimpleServlet.java。

8. 保存目录一样

将生成的 class 文件保存在"C:\jswdk-1.0.1\examples\Web-inf\servlets\"目录中。

9. 运行

对以上程序加以运行。

(1) 在浏览器中输入下面的 URL 地址。

http://localhost:8080/examples/servlet/SimpleServlet?userName=aaa。

(2) 在浏览器中可以看到输出的结果如下所示。

  Say hello to Java Servlet Programming, aaa

  number: 1

(3) 如果再次访问该 Servlet,返回的结果中的第二句会变成如下所示。

  number: 2

四、思考题

(1) 在"C:\JavaWebServer1.1\system\doc"目录下有 Java Web Server 的 HTML 格式的随机文档。API 文档被安装在"C:\JavaWebServer1.1\system\doc\apidoc\pachages.html"中。请自行阅读相关示例和文件,进一步掌握相关知识。

(2) 在"C：\jswdk－1.0.1\examples"目录中有大量可供修改的示例。在"C：\jswdk－1.0.1\"目录中，有 FAQ.html 和 README.html 两个文件。请自行阅读相关示例和文件，进一步掌握相关知识。

(3) 如何将 Servlet/JSP 技术与 JDBC 技术以及 Applet 结合起来，编写 Web 数据库应用程序？

实验十四　综合实验：简单的游戏五子棋

一、开发环境（实验编译及测试环境）

（一）硬件环境

CPU：Intel 奔腾双核 E5200，主频 2.5GHz。

内存：2G。

（二）软件环境

操作系统：Windows 7lipse SDK。

编程语言：Java。

编程环境：JDK7.0。

开发工具：Eclipse。

运行环境：Windows XP 或 Windows 2000 以上操作系统。

二、系统分析

（一）设计的目的

本系统可以实现电脑自动下棋，扫描整个棋盘记录连在一起的黑白棋子数，实现人与电脑有次序地下棋，判断人与电脑的胜负，为电脑下棋提供帮助。

(二)系统包含的类及类之间的关系

本系统共包6个Java源文件。类之间的关系如图14-1所示。

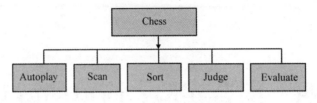

图14-1 类之间的关系

(三)Java源文件及其功能

Java源文件及其功能如表14-1所示。

表14-1 Java源文件及其功能

序号	文件名	主要功能	备注
1	Chess	本程序的主函数	初始化棋盘
2	Autoplay	实现电脑自动下棋	分为8种情况
3	Scan	扫描整个棋盘记录连在一起的黑白棋子数	为判断胜负服务
4	Sort	实现人与电脑有次序地下棋	黑棋白棋有序地下
5	Judge	判断人与电脑的胜负	可以在文本框中显示胜负结果
6	Evaluate	电脑估计	为电脑下棋提供帮助

(四)项目构建思路

此程序旨在打造一个小型五子棋娱乐游戏,经过认真分析和探索,实现小型五子棋游戏的相关功能。

五子棋是一种互动性的益智游戏,需要棋盘、棋子、棋手、裁判各个配置。

1. 首先绘制一个五子棋棋盘

关键代码如下。

```
//绘制棋盘函数
public void paint(Graphics g){
    for (int i=40;i<=400;i=i+20){
        g.drawLine(40,i,400,i);
    }//绘制棋盘行
```

综合实验：简单的游戏五子棋 实验十四

```
        for(int j=40;j<=400;j=j+20){
            g.drawLine(j,40,j,400);
        }//绘制棋盘 列
        g.fillOval(97,97,6,6);    //使用当前颜色填充外接指定矩形框的椭圆
        g.fillOval(337,97,6,6);
        g.fillOval(97,337,6,6);
        g.fillOval(337,337,6,6);
        g.fillOval(217,217,6,6);
    }
```

2. 需要生成两种颜色不一样的棋子

(1) 黑色棋子代码如下。

```
//ChessPoint_black 类继承 Canvas 类
class ChessPoint_black extends Canvas implements MouseListener{
    ChessPadchesspad=null;
    ChessPoint_black(ChessPad p){
        setSize(20,20);//设置棋盘每一小格的长和宽
        addMouseListener(this);
        chesspad=p;
    }
    public void paint(Graphics g){
        g.setColor(Color.black);    //设置棋子颜色为黑色
        g.fillOval(0,0,18,18);//使用黑色填充椭圆棋子
    }
```

(2) 白色棋子代码如下。

```
class ChessPoint_white extends Canvas implements MouseListener{
    ChessPadchesspad=null;
    ChessPoint_white(ChessPad p){
        setSize(20,20);
        addMouseListener(this);
        chesspad=p;
    }
```

```
public void paint(Graphics g){    //绘制棋子的颜色
    g.setColor(Color.white);//设置为白色
    g.fillOval(0,0,18,18);//填充白子
  }
}
```

3. 电脑自动下棋函数设置

为实现双方行棋,我们实现了电脑自动下棋的函数,一共分为 8 种情况,关键代码如下。

```
class AutoPlay{    //AutoPlay 类实现电脑可以自动下棋
  int x,y;  //x 统计玩家的连起来的棋子数
  //y 统计电脑玩家的连起来的棋子数
  void autoPlay(int chesspad[][],int a,int b){
    int randomNumber=(int)(Math.random()*8)+1;  //获取一个随机数
    switch(randomNumber){  //分情况实现电脑自动下棋
      case(1):    //通过 a,b 的值执行不同的运算,最后返回不同的
        //值,达到执行不同的操作
        if(chesspad[a-1][b-1]==0)
            {x=a-1;y=b-1;}
        if(chesspad[a-1][b+1]==0)
            {x=a-1;y=b+1;}
        else if(chesspad[a-2][b-2]==0)
            {x=a-2;y=b-2;}
        else {x=a-3;y=b-3;}
        break;
      case(2):
        if(chesspad[a-1][b]==0)
            {x=a-1;y=b;}
        else if(chesspad[a-2][b]==0)
            {x=a-2;y=b;}
        else {x=a-3;y=b;}
        break;
```

```
case(3):
    if(chesspad[a-1][b+1]==0)
            {x=a-1;y=b+1;}
    else if(chesspad[a-2][b+2]==0)
            {x=a-2;y=b+2;}
    else {x=a-3;y=b+3;}
    break;
case(4):
    if(chesspad[a][b+1]==0)
            {x=a;y=b+1;}
    else if(chesspad[a][b+2]==0)
            {x=a;y=b+2;}
    else {x=a;y=b+3;}
    break;
case(5):
    if(chesspad[a+1][b+1]==0)
            {x=a+1;y=b+1;}
    else if(chesspad[a+2][b+2]==0)
            {x=a+2;y=b+2;}
    else {x=a+3;y=b+3;}
    break;
case(6):
    if(chesspad[a+1][b]==0)
            {x=a+1;y=b;}
    else if(chesspad[a+2][b]==0)
            {x=a+2;y=b;}
    else {x=a+3;y=b;}
    break;
case(7):
    if(chesspad[a+1][b-1]==0)
            {x=a+1;y=b-1;}
    else if(chesspad[a+2][b-2]==0)
```

```
                    {x=a+2;y=b-2;}
            else {x=a+3;y=b-3;}
            break;
        case(8):
            if(chesspad[a][b-1]==0)
                    {x=a;y=b-1;}
            else if(chesspad[a][b-2]==0)
                    {x=a;y=b-2;}
            else{x=a;y=b+3;}
            break;
        }
    }
}
```

4. 规范双方行棋的先后顺序

我们实现了一个判断函数，关键代码如下。

```
class Sort{    //sort 类 实现人与电脑有次序地下棋
    public void sort(int shape[][][]){
        int temp;
        for(int i=0;i<19;i++){
            for(int j=0;j<19;j++){
                for(int h=1;h<=4;h++){
                    for(int w=3;w>=h;w--){
                        if(shape[i][j][w-1]<shape[i][j][w]){
                            //如果前者小于后者,则交换
                            temp=shape[i][j][w-1];
                            shape[i][j][w-1]=shape[i][j][w];
                            shape[i][j][w]=temp;
                        }
                    }
                }
            }
```

    }
}

5. 行棋前的判断

脑只是简单地执行代码，没有智能思考的能力。在行棋之前，它需要对整个棋盘上的棋子做一个判断，然后决定该怎么行棋，关键代码如下。

```java
class Evaluate{  //Evaluate 类
    int max_x,max_y,max;
    public void evaluate(int shape[][][]){
        int i=0,j=0;
        for(i=0;i<19;i++)
        for(j=0;j<19;j++){
            switch(shape[i][j][0]) {
            //电脑根据传进来的三维数组决定该怎么行棋，
                case 5:
                    shape[i][j][4]=200;
                    break;
                case 4:
                    switch(shape[i][j][1]){
                        case 4:
                            shape[i][j][4]=150+shape[i][j][2]+shape[i][j][3];
                            break;
                        case 3:
                            shape[i][j][4]=100+shape[i][j][2]+ shape[i][j][3];
                            break;
                        default:
                            shape[i][j][4]=50+shape[i][j][2]+ shape[i][j][3];
                    }
                    break;
                case 3:
                    switch(shape[i][j][1]){
                        case 3:
```

```
                shape[i][j][4]=75+shape[i][j][2]+shape[i][j][3];
            break;
          default:
                shape[i][j][4]=20+shape[i][j][2]+shape[i][j][3];
        }
          break;
        case 2:
          shape[i][j][4]=10+shape[i][j][1]+shape[i][j][2]+shape[i][j][3];
          break;
        case 1:
          shape[i][j][4]=shape[i][j][0]+shape[i][j][1]+shape[i][j][2]+shape[i][j][3];
          default : shape[i][j][4]=0;
        }
      }
    int x=0,y=0;
    max=0;
    for(x=0;x<19;x++)
      for(y=0;y<19;y++)
        if(max<shape[x][y][4]){
        max=shape[x][y][4];
        max_x=x;   max_y=y;}
  }
}
```

6. 在双方行棋后,我们需要判断哪一方赢

裁判的功能代码如下。

```
//判断人与电脑的胜负
class Judge{
  static boolean judge(int a[][],int color){
    int i,j,flag;
    for(i=0;i<19;i++){    //行
```

```
flag=0;
for(j=0;j<19;j++)//列
   if(a[i][j]==color){
      flag++;  //棋子计数器
      if (flag==5)//如果棋子数等于5
      return true;}  //返回true
      else  flag=0;  //返回false


}


for(j=0;j<19;j++){
   flag=0;  //棋子计数器
   for(i=0;i<19;i++)
   if(a[i][j]==color) //每一列的棋子颜色一样
   {
       flag++;
   if(flag==5)   //棋子数为5
     return true;}  //返回true 否则返回true
   else flag=0;
}


for(j=4;j<19;j++){
flag=0;
int m=j;
for(i=0;i<=j;i++){
   if(a[i][m--]==color){  //对角线棋字数一样
      flag++;
      if(flag==5)  //为5则返回true
      return true;}
      else flag=0;}  //棋子计数器归零
}
```

```
for(j=14;j>=0;j--){
    flag=0;  int m=j;
    for(i=0;i<=18-j;i++){

        if(a[i][m++]==color){  //对角线棋子数是否一样
            flag++;
            if(flag==5)
            return true;}  //棋子书为5则返回true
            else flag=0;}  //否则归零
}


for(i=14;i>=0;i--){
    flag=0;  int n=i;
    for(j=0;j<19-i;j++){

        if(a[n++][j]==color){
            flag++;
            if(flag==5)
            return true;}
            else flag=0;}
}


for(j=14;j>=0;j--){
    flag=0; int m=j;
    for(i=18;i>=j;i--){
        if(a[i][m++]==color){
            flag++;
            if(flag==5)
            return true;}
            else flag=0;}
}
```

return false;}
}

7. 程序的可视化

我们最终的程序是基于可视化的,所以需要用容器及组建来实现,关键代码如下。

```
class ChessPad extends Panel implements MouseListener,ActionListener{
//创建棋盘
    int array[][]=new int[19][19];//创建二维数组,生成棋盘


    Sort sort=new Sort();
    int i=0;   //控制棋子颜色
    int x=-1,y=-1,棋子颜色=1;
      Button button=new Button("重新开局");    //确定按钮
    TextField text_1=new TextField("请黑棋下子"),//创建文本组件 并初始化第一个
        text_2=new TextField(),
        text_3=new TextField();
class ChessPad extends Panel implements MouseListener,ActionListener{
//创建棋盘
    int array[][]=new int[19][19];//创建二维数组,生成棋盘


    Sort sort=new Sort();
    int i=0;   //控制棋子颜色
    int x=-1,y=-1,棋子颜色=1;
      Button button=new Button("重新开局");    //确定按钮
    TextField text_1=new TextField("请黑棋下子"),//创建文本组件 并初始化第一个
        text_2=new TextField(),
        text_3=new TextField();


    ChessPad(){   //ChessPad 函数的构造函数
      setSize(440,440);   //
      setLayout(null);
```

setBackground(Color.pink);//设置背景为粉色

addMouseListener(this);

add(button);//向组件添加指定的重新开局菜单

button.setBounds(10,5,60,26);

//移动组件并调整其大小。由 x 和 y 指定左上角的新位置,由 width 和 height 指定新的大小。

button.addActionListener(this); //添加指定的操作侦听器,以接收来自此按钮的操作事件

　　add(text_1);　text_1.setBounds(90,5,90,24);

　　add(text_2);　text_2.setBounds(290,5,90,24);

　　add(text_3);　text_3.setBounds(200,5,80,24);

　　for(int i=0;i<19;i++)

　　　for(int j=0;j<19;j++)

　　　　{array[i][j]=0;}　//初始化数组为 0

　　for(int i=0;i<19;i++)

　　　for(int j=0;j<19;j++)

　　　　for(int h=0;h<5;h++)

　　　　　{scanp.shape[i][j][h]=0;

　　　　　scanc.shape[i][j][h]=0;}　//初始化三维数组为 0

　　text_1.setEditable(false);

　　text_2.setEditable(false);　//设置文本组件为不可编辑

}

三、模块功能介绍

(一) 主类 Chess

1. Chess 成员变量

Chess 成员变量如表 14-2 所示。

表 14 - 2 Chess 主要成员变量(属性)

成员变量描述	变量类型	名称
控制棋子颜色	int	i
重新开局	Button	Button
请黑棋下子	TextField	text_1
请白棋下子	TextField	text_2
这是第"+i+"步	TextField	text_3

2. Chess 主要方法

Chess 主要方法如表 14 - 3 所示。

表 14 - 3 Chess 主要方法

方法名称	返回类型	功能	备注
ChessPad()	初始化棋盘	设置棋盘颜色	ChessPad()
Void paint (Graphics g)	对棋子等图像进行初始化	设置棋子大小、颜色等	Void paint (Graphics g)
Public void mousePressed (MouseEvent e)	鼠标的实践监听	单击实践	public void mousePressed (MouseEvent e)
public void paint (Graphics g)	画出图像	控制图像的大小范围	public void paint (Graphics g)
public void mousePressed (MouseEvent e)	处理按下鼠标的事件	选中下棋的位置	public void mousePressed (MouseEvent e)
public void mouseReleased (MouseEvent e)	处理鼠标离开的事件	本步下棋结束	public void mouseReleased (MouseEvent e)
public void mouseExited (MouseEvent e)	处理鼠标离开棋盘时	鼠标离开组件不实现任何事件	public void mouseExited (MouseEvent e)
public void mouseClicked (MouseEvent e)	处理发生单击的事件	实现下棋以及重新开始游戏	public void mouseClicked (MouseEvent e)

源代码见文件 Chess.java 中的公共类 Chess。

(二) 类 Autoplay

1. Autoplay 成员变量

Autoplay 成员变量如表 14 - 4 所示。

表 14-4　Autoplay 主要成员变量

成员变量描述	变量类型	名称
统计玩家连起来的棋子数	int	X
电脑玩家连起来的棋子数	Int	Y

2. Autoplay 主要方法

Autoplay 主要方法见表 14-5 所示。

表 14-5　Autoplay 主要方法

方法名	功能	备注
void autoPlay（int chesspad[ ][ ],int a,int b)	分情况实现电脑自动下棋	使用 switch 语句分为 8 种情况来下棋

源代码见文件 Chess.java 中的 Autoplay 类。

(三) 类 Scan

1. Scan 主要成员变量

Scan 主要成员变量如表 14-6 所示。

表 2-5　Scan 主要成员变量

成员变量描述	变量类型	名称
统计棋盘种五个连起来的棋子数	数组	int shape[ ][ ][ ]
控制行数	int	i
控制列数	int	j

2. Scan 主要方法

Scan 主要方法如表 14-7 所示。

表 14-7　Scan 主要方法

方法名	功能	备注
void scan（int chesspad[ ][ ],int colour)｛	判断棋盘上连在一起的黑/白棋子个数	

源代码见文件 Chess.java 中的 Scan 类。

四、功能测试及运行效果

（一）系统主界面

程序运行的主要界面如图 14-2 至图 14-4 所示。

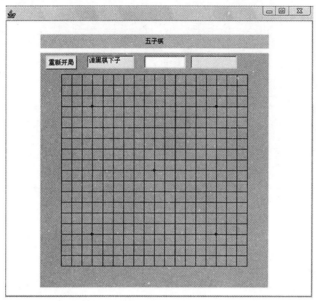

图 14-2 初始化棋盘效果，玩家为黑棋

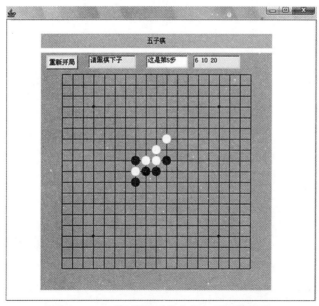

图 14-3 可以统计下棋步数与下棋的位置

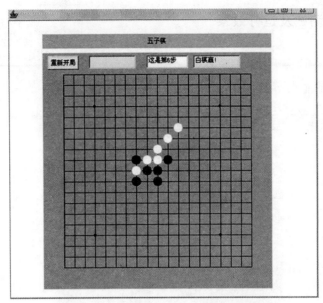

图 14-4 可以实现胜负的判断并终止游戏

五、思考题

（1）图形绘制的过程中常用的布局管理器有哪些？

（2）如何改善界面的外观？

参考文献

[1] 辛运帏,饶一梅.Java 语言程序设计(第 2 版)[M].北京:人民邮电出版社,2015.

[2] 叶乃文,王丹.Java 语言程序设计教程[M].北京:机械工业出版社,2010.

[3] 黄岚,王岩,王康平.Java 程序设计(第 2 版)[M].机械工业出版社,2016.

互联网+教育+出版

教育信息化趋势下，课堂教学的创新催生教材的创新，互联网+教育的融合创新，教材呈现全新的表现形式——教材即课堂。

立方书

轻松备课　分享资源　发送通知　作业评测　互动讨论

"一本书"带走"一个课堂"　　教学改革从"扫一扫"开始

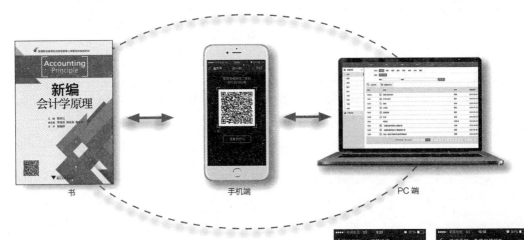

书　　　　　　手机端　　　　　　PC端

打造中国大学课堂新模式

【创新的教学体验】
开课教师可免费申请"立方书"开课，利用本书配套的资源及自己上传的资源进行教学。

【方便的班级管理】
教师可以轻松创建、管理自己的课堂，后台控制简便，可视化操作，一体化管理。

【完善的教学功能】
课程模块、资源内容随心排列，备课、开课、管理学生、发送通知、分享资源、布置和批改作业、组织讨论答疑、开展教学互动。

扫一扫 下载APP

教师开课流程
→ 在APP内扫描封面二维码，申请资源
→ 开通教师权限，登录网站
→ 创建课堂，生成课堂二维码
→ 学生扫码加入课堂，轻松上课

网站地址：www.lifangshu.com
技术支持：lifangshu2015@126.com；电话：0571-88273329